Pesticide residues in food - 1983

Report of the joint meeting
of the
FAO Panel of Experts on Pesticide Residues
in Food and the Environment
and the
WHO Expert Group on Pesticide Residues
held in Geneva
5-14 December 1983

FAO
PLANT
PRODUCTION
AND PROTECTION
PAPER
56

FOOD
AND
AGRICULTURE
ORGANIZATION
OF THE
UNITED NATIONS
Rome, 1984

M-84

ISBN 92-5-102094-9

CONTENTS

---

* Compounds re-evaluated by the present meeting.

1983 JOINT MEETING OF THE FAO PANEL OF EXPERTS ON PESTICIDE RESIDUES IN FOOD AND THE ENVIRONMENT AND THE WHO EXPERT GROUP ON PESTICIDE RESIDUES

Geneva, 5-14 December 1983

FAO Panel of Experts on Pesticide Residues in Food and the Environment

A. Ambrus, Plant Protection and Agrochemistry Centre, Budapest, Hungary

J.A.R. Bates, Pesticide Registration Department, Ministry of Agriculture, Fisheries and Food, Hatching Green, Harpenden, Herts, UK (Rapporteur)

K.R. Hill, Chemistry Laboratory, US Department of Agriculture, Agricultural Environment Quality Institute, Beltsville, Md, USA (Chairman)

N.F. Ives, Office of Pesticide Programs, Environmental Protection Agency, Washington DC, USA

J.T. Snelson, Department of Primary Industry, Canberra, Australia(Vice-Chairman)

K. Voldum-Clausen, Division of Pesticides and Contaminants, National Food Institute, Soborg, Denmark

Observer invited by FAO

A.J. Pieters, Directorate of Public Health, Foodstuffs Division, Leidschendam, The Netherlands (Chairman of the Codex Committee on Pesticide Residues)

WHO Expert Group on Pesticide Residues

V. Benes, Department of Toxicology, Institute of Hygiene and Epidemiology, Prague, Czechoslovakia

D. Clegg, Pesticide Section, Toxicological Evaluation Division, Food Directorate, Health Protection Branch, Ottawa, Canada (Chairman)

S. Jetanasen*, Epidemiology Division, Office of the Under-Secretary, Ministry of Public Health, Bangkok, Thailand

M. Lotti, Istituto di Medicina del Lavoro, Padua, Italy (Rapporteur)

D. de Paola, Department of Pathology, School of Medicine, Federal University, Rio de Janeiro, Brazil

O.E. Paynter, Hazard Evaluation Division, US Environmental Protection Agency, Washington DC, USA

---

* Invited but unable to attend.

FAO/WHO Secretariat

A.L. Black, Commonwealth Department of Health, Woden, Australia (WHO Temporary Adviser)

J.R. Cabral, Unit of Mechanisms of Carcinogenesis, International Agency for Research on Cancer, Lyon, France

A. di Muccio, Toxicological Laboratory - Pesticide Section, Istituto Superiore di Sanità, Rome, Italy (WHO Temporary Adviser)

R.B. Jaeger, Toxicology Branch, Hazard Evaluation Division, US Environmental Protection Agency, Washington DC, USA (WHO Temporary Adviser)

F-W. Kopisch-Obuch, Plant Protection Service, FAO, Rome, Italy (Secretary of FAO Panel of Experts on Pesticide Residues in Food and the Environment)

L.G. Ladomery, Joint FAO/WHO Food Standards Programme, FAO, Rome, Italy

A.F. Machin, Ministry of Agriculture, Fisheries and Food, Central Veterinary Laboratory, Weybridge, Surrey,UK (FAO Consultant)

T. Pekkanen, Department of Food Hygiene, College of Veterinary Medicine, Helsinki, Finland (WHO Temporary Adviser)

M. Repetto*, Regional Toxicology Institute, Seville, Spain (WHO Temporary Adviser)

M. Tobe, Toxicological Division, National Institute of Hygienic Sciences, Tokyo, Japan (WHO Temporary Adviser)

G. Vettorazzi, International Programme on Chemical Safety, Division of Environmental Health, WHO, Geneva, Switzerland (Secretary of the WHO Expert Group on Pesticide Residues)

E. Weisenberg, Institute for the Standardization and Control of Pharmaceuticals, Ministry of Health, Jerusalem, Israel (WHO Temporary Adviser)

D.L. Wu-Lau*, Pesticides Section, Toxicology Evaluation Division, Food Directorate, Health Protection Branch, Ottawa, Canada (WHO Temporary Adviser)

Additional WHO Temporary Advisers**

E.R. Budd, Office of Pesticide Programs, US Environmental Protection Agency, Washington DC, USA

N. Ito, First Department of Pathology, Nagoya City University, Medical School, Nagoya, Japan

P. Shubik, Green College at the Radcliffe Observatory, Oxford OX2 6HG,UK

* Invited but unable to attend

** Present only at the discussion of o-phenylphenol and its sodium salt.

PESTICIDE RESIDUES IN FOOD

Report of the 1983 Joint FAO/WHO Meeting of Experts

1. INTRODUCTION

A Joint Meeting of the FAO Panel of Experts on Pesticide Residues in Food and the Environment and the WHO Expert Group on Pesticide Residues (JMPR) was held in Geneva, Switzerland, from 5-14 December 1983. The meeting was opened by Dr Mercier, International Programme on Chemical Safety, WHO, on behalf of the Directors-General of FAO and WHO. The FAO Panel had already met in preparatory sessions on 1 and 2 December.

The meeting was held in pursuance of recommendations made by previous meetings and accepted by the governing bodies of FAO and WHO that studies should be undertaken jointly by experts to evaluate possible hazards to man arising from the occurrence of residues of pesticides in foods. The reports of previous joint meetings (see references on p. 48) contain information on acceptable daily intakes for man (ADIs), maximum residue limits (MRLs) and general principles of evaluation for the various pesticides considered. The supporting documents contain detailed monographs on these pesticides and include comments on analytical methods. The present meeting was convened to consider a further number of pesticides together with items of a general or a specific nature. These include items for clarification of recommendations made at previous meetings or for reconsideration of previous evaluations in the light of findings of subsequent research or other developments.

During the meeting the FAO Panel of Experts was primarily responsible for reviewing pesticide use patterns (good agricultural practice) data on the chemistry, composition and methods of analysis of the pesticides and for estimating the maximum residue levels that may occur following the use of the pesticide according to good agricultural practice (GAP). The WHO Expert Group was primarily responsible for reviewing toxicological and related data and for estimating, where possible, ADIs for man for the pesticides. The recommendations of the Joint Meeting, including those for ADIs and maximum residue limits and the indications for further research and information, are proposed for use by Member Governments of the respective Agencies and other interested parties.

2. GENERAL CONSIDERATIONS

2.1 MODIFICATIONS TO THE AGENDA

The agenda was modified to permit consideration of amitraz, azocyclotin, chinomethionat, cyhexatin, cypermethrin, diflubenzuron and ethiocarb (in response to questions from the Codex Committee on Pesticide Residues (CCPR)), dithiocarbamates, fenitrothion, permethrin and 2-phenylphenol. The definition and expression of the residue was reviewed for the following compounds not already on the agenda for other reasons: azinphos-ethyl, carbophenothion, ethion, fenchlorphos, fensulfothion, fenthion and propoxur.

Isoprocarb and terbufos, recommended for priority attention by the 14th session of the CCPR, and fenarimol, recommended for priority attention by the

13th session of the CCPR and postponed by the 1982 JMPR, were not considered because no data had been submitted in time. All three compounds were transferred to draft agenda for future meetings (see section 6). Cyanofenphos was deleted from the agenda on receipt of information that it was no longer being manufactured. Reviews on phorate, due for review in 1983 (see 1982 Report), methacrifos (see 1980 Report) and methidathion were postponed owing to lack of data.

2.2 GENERAL PRINCIPLES FOR ESTIMATING ADIs

The meeting took account of the principles elaborated in the reports of previous meetings concerning the estimation of ADIs and the formulation of other decisions based on toxicological considerations. It also took account of a number of reports of other committees,in particular the principles and guidelines contained in the Seventeenth Report of the Joint FAO/WHO Expert Committee on Food Additives,[1/] the report of the WHO Scientific Group on Procedures for Investigating Intentional and Unintentional Food Additives[2/] and the report of the WHO Scientific Group on Assessment of the Carcinogenicity and Mutagenicity of Chemicals[3/]. In reaffirming its general adherence to the principles laid down previously in estimating ADIs or temporary ADIs for pesticides, the meeting felt that some of those principles should be clarified, rephrased or extended.

2.2.1 Methodology

The meeting also endorsed recommendation 4 of the Joint FAO/WHO Expert Committee on Food Additives (JECFA),[4/] which states:

"The multidisciplinary approach to toxicology in recent years has led to significant advances in methodological approaches and procedures for evaluation. Paradoxically, these advances have sometimes made it more difficult to arrive at clear recommendations concerning food additives and contaminants. Examples of problems that have been encountered in the general context of toxicological testing include:

(a) the influence of nutritional imbalance on the manifestation of toxic effects including carcinogenic effects;

(b) the interpretation of results obtained when excessively high doses are used; and

(c) the testing of foods for the presence of small amounts of toxicologically active components.

[1/] WHO, Technical Report Series, No 539; FAO Nutrition Meetings Report Series, No 53, 1974.

[2/] WHO, Technical Report Series, No 348, 1967

[3/] WHO, Technical Report Series, No 546, 1974

[4/] WHO, Technical Report Series, No 669, 1981

Examples of problems encountered in testing for carcinogenic potential include:

(a) the interpretation of enhancement of some types of spontaneously occurring tumours in susceptible strains of certain animal species;

(b) the implications for toxicological evaluations of promoting factors and other modifiers of carcinogenic activity; and

(c) the relationship of results of various kinds of mutagenicity tests to the prediction of carcinogenic potential.

In order to resolve these problems, the Committee strongly recommended that a group of experts be convened, as soon as possible, to study the application of advances in methodology to the toxicological evaluation of food additives and contaminants, and also of pesticide residues."

The meeting was supplied with two new documents, a background paper describing past recommendations of both JMPR and JECFA on this subject from the Secretariat.[5/], and the report of a Strategy Meeting held in Oxford, UK, 19-23 September 1983.[6/] The meeting noted that this Strategy Meeting defined its objectives as providing the JECFA with specific recommendations on updating: (a) the principles enunciated in JECFA reports concerning safety evaluation in relation to specific toxicological problems or to specific chemical entities or groups; (b) the test methods used in the toxicological evaluation of chemicals in food, and (c) the assessment procedures adopted by the JECFA in determining quantitative end points.

The meeting further noted that the Strategy Meeting recognized the differentiation introduced between products which should exert no toxic manifestations in use and those whose use depends on their toxicity. The former include deliberate food additives; the latter, compounds used as pesticides. The goal of the evaluation of the first group is to assess their potential toxicity to assure safety in use, whereas for the second group it is to quantify their toxicity to establish their safety. It should be recognized that the levels of pesticide residues in food under consideration are normally low and frequently close to the limits of analytical sensitivity.

The meeting considered there was a need to undertake a parallel exercise for the toxicological assessment of pesticides leaving residues in food. However, it was agreed that it would be advisable to await the views of the JECFA on this subject and then to study the applicability of their conclusions to pesticide residues before undertaking a separate review of methodology.

2.2.2 No-observed-effect level (NOEL)

The meeting considered terminology for the expression of the level found to cause no significant toxicological effect in a study. The meeting agreed

---

5/"Updating Methodology for Testing and Assessing Chemicals in Food", WHO Secretariat, 1983

6/International Programme on Chemical Safety, Report of a Strategy Meeting on Updating Principles of Methodology for Testing and Assessing Chemicals in Food, Oxford, UK, 19-23 September 1983.

that the term "no-observed-effect level" reflected the evaluation process more accurately than the term "no-effect level". The meeting recommended the use of the former term, abbreviated to NOEL, as future JMPR terminology.

2.2.3 Definition of ADI

The meeting was informed that one statement about general principles for estimating ADIs in the 1982 JMPR report had generated a certain confusion. This meeting wishes to make it clear that the definition of the ADI in the report of the 1975 JMPR is still valid and that the term "man" refers to mankind (*Homo sapiens*), although it might not always be applicable to special groups such as chronically ill persons, pregnant women or children. Studies relevant to the assessment of possible risks to such groups are normally considered in the evaluation of new pesticides and may become available for older compounds.

2.3 EXPRESSION OF MAXIMUM RESIDUE LIMITS IN TERMS OF A SINGLE COMPOUND

The re-evaluation of some residue descriptions outlined in para 3.3 of this report presented the meeting with the opportunity to discuss the concept, use and definition of MRLs in relation to the overall evaluation of the safety of pesticide residues in food. It was agreed that the primary goal of the evaluation of residues data is to estimate the maximum residue levels of a pesticide and its metabolites which may occur in food when the pesticide is used according to good agricultural practice in order that the residue may be assessed in relation to the known toxicity of the compound. Following this approach, the composition and the level of the total residue of toxicological significance have to be determined by metabolic studies and supervised residues trials to establish the acceptability of the proposed use of the pesticide. One way in which this acceptability is expressed is by the use of the estimated maximum residue level as a legal limit, referred to as the maximum residue limit or MRL.

The monitoring of good agricultural practice can be reasonably achieved by analysing food commodities and comparing the results with the appropriate MRL. The description of the residue to be measured is of course of paramount importance and,when the total residue is toxicologically acceptable, it may be possible to identify a single chemical compound, e.g. the parent compound, a metabolite or a derivative produced in an analytical procedure, the measurement of which can be considered to indicate the total significant residue.

There are appreciable advantages in the above approach, since not only can more laboratories participate in regulatory control (since a relatively simple and rapid analytical procedure may not require the expensive equipment and time required for an extensive determination of all components of a residue) but the overall cost can be reduced and many more samples may be analysed by the regulatory laboratories.

For the selection of a single compound to represent the residue the following criteria have been used by the meeting:

1. The concentration of the compound should bear a known relationship to the concentration of the toxicologically significant residues in the sample.

2. The compound should be:

   i. sufficiently stable to allow reproducible analysis

   ii. available as an analytical standard

   iii. as far as possible, recoverable in multi-residue procedures.

The meeting agreed that, whenever possible, it would continue to follow the principle of expressing limits in terms of a single compound. Data on the total residue and its components under practical conditions of use will continue to be required and will contribute to the establishment of the relationship of an indicator compound to the total residue. Consequently, manufacturers and governments submitting residues data for evaluation are urged to present their results in a form that will enable the JMPR to consider whether a single component is suitable for defining the MRL.

It is proposed that all relevant data taken into account both in reaching a conclusion on the acceptability of the proposed use of the pesticide and the selection of the indicator compound should continue to be reported in the records of the meeting.

It should be noted however that, regardless of the definition of the residue to which MRLs apply, total diet studies and similar dietary investigations should determine the total amount of the toxicologically significant residue in order to achieve a realistic estimate of pesticide intake in food and a base-line for comparison with ADIs, if desired on a national basis.

2.4 EVALUATION OF CARCINOGENICITY DATA

The meeting recognized changing concepts governing practical approaches to the control of chemical carcinogens. While it is widely recognized that chemical carcinogens act via different mechanisms, none of these mechanisms is fully understood. Therefore no hypothesis can be considered to be entirely free from possible fault. However, in certain instances a chain of events has been associated with carcinogenesis and the interruption of one event can obviously lead to an interruption of the final outcome. Thus compounds leading to the formation of bladder calculi in rodents may in turn lead to bladder tumours. In the absence of calculi,no tumours occur. Similarly, certain hormones give rise to tumours following excessive hormonal stimulation. Again, if there is no hormonal stimulation, no tumours will occur. Thus cancer can be prevented without complete knowledge of the mechanism of carcinogenesis. Other examples of similar mechanisms must be sought to provide a basis for scientifically establishing tolerance levels. An understanding of metabolic processes and their relationship to carcinogenesis clearly needs to be emphasized.

Two approaches have been given much attention in recent years. These are (1) the use of mathematical models for risk assessment and (2) the possibility of classifying carcinogens as "genotoxic" or "epigenetic" ("non-genotoxic"). In so far as mathematical models are concerned it has become apparent that different chemical carcinogens for which adequate dose-response data are available (and there are all too few of these) fit different mathematical models. This finding has been a major basis for the concern that no tolerance level for carcinogens can exist. However, a problem that arises is that many carcinogens yield data that can be fitted by several different mathematical models. Unfortunately, these different models (multi-hit, Weibull, for example) can yield results that determine NOELs varying by as much as 10^3 or more. Thus, from a practical point of view, mathematical modelling has indicated that (a) a NOEL for some carcinogens is likely to exist but (b) NOELs are impossible to establish with scientific certainty. The use of arbitrary safety factors would, therefore, seem to be a reasonable approach.

The terms "genotoxic" and "epigenetic" have been used for grouping carcinogens in terms of their mechanism of action. "Genotoxic" has been applied to those agents that modify the DNA structure. Agents that are found to be carcinogenic in animal experiments but show no interaction with DNA have been called epigenetic (non-genotoxic) carcinogens. This classification has arisen as a result of the increased use of short-term *in vitro* tests for mutagenicity of chemicals. As more data have accumulated it has become apparent that by no means all mutagens are carcinogenic or *vice versa*. However, it is the case that many of the so-called classic potent carcinogens are often chemically reactive, either as alkylating agents or by otherwise affecting DNA. Even though it is premature to classify carcinogens into simple categories on this basis it is, nonetheless, important to take into account all aspects of the activities of such agents in arriving at a risk assessment.

A risk assessment of a carcinogen should, if possible, be based on the evaluation of all available data and not on a single bioassay. Certain end-points in carcinogenicity studies, such as the enhancement of one or more of the common spontaneous tumours of rodents, may carry less weight with the toxicologist than other findings. The enhancement of the occurrence of hepatomas, lung adenomas, lymphomas and other tumours, for example, are frequently reported. It is most important, in the first instance, to evaluate the robustness of the data. Among other issues it is essential to relate concurrent control incidence levels to historical control knowledge. If concurrent controls, for example, deviate considerably from known historical levels, then reasons for this must be sought.

There is little knowledge of the etiological factors concerned in the origin of most rodent spontaneous tumours. Notable exceptions are virally mediated lymphomas, some mammary tumours and scattered other lesions. No such knowledge is available for the common hepatomas and lung adenomas. It is, therefore, difficult to extrapolate to man findings that are concerned only with the enhancement of such lesions.

However, the enhancement of these spontaneous lesions has been found to correlate to some extent with other forms of carcinogenicity (namely induction of tumours not seen, or only rarely seen, in controls). The enhancement of these common spontaneous lesions must therefore be taken to be an indication of possible carcinogenicity, which must be resolved by further experiments.

2.5 LATENCY PERIOD

The meeting noted that one of the recommended criteria for the asessment of carcinogenicity, the latency period, can rarely be utilized because of lack of suitable information on the time taken for tumours to appear.

The meeting was of the opinion that this undesirable state of affairs could be alleviated, to some degree, by modifications of existing protocols on carcinogenicity studies to include small satellite groups of animals, which could then be sacrificed at intervals prior to the termination of the study. Such satellite groups could be limited to control and top dosage groups. In certain cases (e.g. where liver or thyroid hyperplasia occurs), careful assessment of 90-day studies can be of value in determining the necessity for, and the frequency of, sacrifice of such satellite groups of animals.

Further, with modern advances in computer techniques, the accuracy of morphometric measurements has markedly increased in recent years. The data on morphometric measurement of tumours should be further investigated to determine its usefulness in assessing the latency period in standard-protocol carcinogenicity studies.

2.6 RESULTS OF MUTAGENICITY TESTS

The meeting considered the implications of the results of mutagenicity studies and their impact on the evaluation of ADIs. In the present state of knowledge, the use of mutagenicity data derived from tests often performed in bacterial or fungal systems *in vitro* or in a limited range of *in vivo* procedures in experimental animals is believed to be useful as a screen for potential carcinogenic activity, although the meeting recognized that the association between mutagenic and carcinogenic activity is uncertain. However, where suitable long-term carcinogenicity studies are available, the results from these studies must be considered to override possible concerns raised by mutagenicity data.

The meeting, in considering mutagenicity tests *per se*, was unable to determine the relevance of the results to possible human health hazards. Hence, such data cannot be utilized directly in the assessment of the ADI.

Overall, at this time, mutagenicity data can therefore only be utilized as ancilliary data in the deliberations of the meeting. For this reason consideration was given to the reporting of such data in the monographs. The considerable numbers of mutagenicity studies being reported on individual pesticides results in a disproportionate amount of the monograph being devoted to such studies in relation to their importance in the meeting's deliberations. Given that the detailed data on mutagenicity studies have been adequately reviewed by the meeting and have been accepted as valid, it is recommended that future meetings should report these data only in tabular form, thus de-emphasizing the apparent importance of the role of these studies in assessing the ADI to a more realistic level.

Summary tables should include:

Genetic endpoint or test system (i.e. micronucleus, DNA repair, etc)
Test object (i.e. *E. coli*, *S. typhimurium*)
Purity of pesticide used
Concentration(s) of pesticide
Results
Reference

2.7 DATA AVAILABILITY AND WITHDRAWAL OF ADIs

The assessment of the safety of a compound is a continuing process. New information may become available and may require a revision of previous decisions. In the current terminology, this situation is referred to as "re-evaluation of a pesticide". This task has frequently required the adoption of administrative procedures designed to ensure the continued awareness of parties interested in generating scientific data and aimed at establishing the safety of pesticides. The adoption of temporary ADIs[7] and/or maximum residue limits, the indication of further work required and/or desirable, and the establishment of deadlines for the submission of this work are a few examples of such administrative measures.

Further work "required" is work that must be done, properly reported and made available, within a specified period before an acceptable daily intake for man and/or maximum residue limits can be recommended or confirmed. In certain instances, although acceptable daily intakes have been established, further work has been considered to be essential to remove doubts about the toxicological significance of some experimental observations. Results of the further work required should be made available not later than the specified date, after which the compound will be re-evaluated. The re-evaluation may be done earlier if relevant information should become available.

In some instances the meeting was unable to estimate an acceptable daily intake, either because adequate data were lacking or because the available information was unsuitable. Furthermore, on several occasions the meeting withdrew existing ADIs or temporary ADIs either because the data required were not made available or because new information showed the need for a revision of previous decisions made on data which might now be regarded as insufficient in the light of more recent information.

The meeting emphasized that such a withdrawal of an ADI or of a temporary ADI does not necessarily indicate a potential health hazard but only that insufficient information is available at the time of review to permit the meeting to state with reasonable certainty that there is no likelihood of adverse effects on health resulting from ingestion over a prolonged period.

The meeting considered the decision of previous meetings with regard to the withdrawal of an ADI as well as the view of the Joint FAO/WHO Expert Committee on Food Additives expressed in its 11th Report[8] and agreed that this procedure should not be interpreted as casting doubt on the safety of the substance, nor should it be considered tantamount to a recommendation for its withdrawal from use.

2.8 REPORT OF THE 15TH SESSION OF THE CODEX COMMITTEE ON PESTICIDE RESIDUES (Alinorm 85/24)

The meeting considered the draft report of the 15th session of the CCPR, in particular those paragraphs in which the CCPR had drawn certain items to the attention of the JMPR.

7/ FAO/WHO, 1973, Pesticide Residues in Food. Report of the 1972 FAO/WHO Meeting on Pesticide Residues. FAO Agricultural Studies, No. 90; WHO Technical Report Series, No 525.

8/ FAO/WHO, 1968, Specifications for the identity and purity of food additives and their toxicological evaluation. 11th Report of the JECFA, FAO Nutrition Meeting Report Series, No.44; WHO Technical Report Series, No.383.

Questions on individual pesticides

Several questions raised at the CCPR on individual pesticides were considered at the meeting and are dealt with in Section 3.1 and the appropriate parts of Section 4.

These were:

- cyhexatin	(paras 116-117)	- see 4.15
- chinomethionat	(para. 124)	- see 4.10
- ethiofencarb	(para. 152)	- see 4.25
- cypermethrin	(para. 165)	- see 4.16
- pyrethroid insecticides	(para. 177)	- see 3.1
- amitraz	(para. 182)	- see 4.1
- azocyclotin	(paras 202-203)	- see 4.3
- diflubenzuron	(para. 205)	- see 4.21

2.9 WORKING LANGUAGE OF THE JOINT MEETING

The working language of the JMPR is English. Translation services are not available during the meeting and considerable problems exist in obtaining translations of material submitted prior to the meeting.

The present meeting was unable to evaluate submitted data on some compounds,as English translations were not available. Therefore, the JMPR requests that submitted data for all compounds should be provided in the working language of the meeting, i.e. English.

3. SPECIFIC PROBLEMS

3.1 REVIEW OF MAXIMUM RESIDUE LIMITS FOR SYNTHETIC PYRETHROIDS AND FENVALERATE

The 15th Session of the CCPR (1983) requested the JMPR to consider the differences in the MRLs proposed for cypermethrin, deltamethrin, permethrin and fenvalerate, some of which seemed anomalous in view of their relative application rates. The meeting considered a working paper which outlined the problem and agreed that a review of these MRLs was appropriate but could not be covered adequately at the present meeting. The meeting proposed that the review be carried out by the 1984 JMPR.

3.2 RE-EVALUATION OF THE DEFINITION OF SOME PESTICIDE RESIDUES TO WHICH MRLs REFER

Acting on the recommendation of the 1982 JMPR (Report, section 2.5),FAO arranged for the re-evaluation of those pesticides identified by the 14th Session of the CCPR as requiring consideration. The principles applied in the re-evaluation of the original data are outlined in section 2.5 of this report.

The following compounds were re-evaluated and the information taken into account in making recommendations for changes to the definitions is summarized in section 4.

azinphos-ethyl	(4.2)
carbophenothion	(4.9)
chlorothalonil	(4.12)
coumaphos	(4.13)
ethion	(4.27)
fenchlorphos	(4.28)
fenitrothion	(4.29)
fensulfothion	(4.30)
fenthion	(4.31)
pirimiphos-methyl	(4.41)
propoxur	(4.43)

As a result of the reviews several definitions were simplified. Since the new definitions do not alter the use patterns of the pesticides, nor the pre-harvest intervals on which the original MRLs or guideline levels were based, the total residue that may reach the consumer also remains the same. Consequently, the changes to the existing recommendations may be considered to be non-substantial.

3.3 MEASUREMENT OF CHOLINESTERASE INHIBITION BY CARBAMATE PESTICIDES

The meeting noted that in the reports of several studies on carbamate pesticides the method of determination of cholinesterase inhibition was inadequately reported, and occasionally data were inconsistent with respect to dose and the degree of cholinesterase inhibition. Carbamates are usually considered to be reversible inhibitors of cholinesterase with a short duration of action. Because of the reversal of inhibition of the enzyme by dilution, as would occur during the preparation of the assay, inhibition cannot be accurately measured. The meeting stressed that,in order to permit evaluation of cholinesterase inhibition by carbamates *in vivo*, special care is required in reporting all details of such studies.

4. EVALUATION OF DATA FOR ACCEPTABLE DAILY INTAKE FOR MAN AND MAXIMUM RESIDUE LIMITS

EXPLANATION

This section contains brief reports on the compounds considered by the present meeting. The information provides a summary of the material that will appear in the full monographs (1983 Evaluations of some pesticide residues in food), including details of further work or information considered necessary or desirable by the meeting. The requirements for further work or information are additional to those mentioned in earlier reports that have not been previously satisfied.

As in previous evaluations, compounds that were re-evaluated by the present meeting are marked with an asterisk. Compounds evaluated for the first time are identified by their chemical names, acording to IUPAC nomenclature, as well as their common names. Standard common names of the International Organisation for Standardization (ISO) are used wherever possible.

The estimated maximum acceptable daily intakes for man (ADI) and maximum residue limits (MRLs), together with guidelines levels (GLs) for compounds with no currently estimated ADI, appear in Annex 1. New MRLs classified according to commodity groups are listed in Annex 2A. Changes in MRLs are listed in Annex 2B.

4.1 AMITRAZ*

Residue and analytical aspects

The meeting reviewed the data on residues in carcase meat of sheep in the 1980 evaluations and concluded that an estimated maximum residue level of 0.1 mg/kg was more appropriate than the figure of 0.2 mg/kg previously estimated. After a re-examination of residue data on cucumbers the meeting re-affirmed the estimate of 0.5 mg/kg as a maximum residue likely to occur in good agricultural practice.

4.2 AZINPHOS-ETHYL*

Residue and analytical aspects

The CCPR asked the JMPR to reconsider the definition of the residue. The residue data summarized in the 1973 Evaluations indicate that the oxygen analogue is not detectable or its concentration is lower than 10% of the total residue. Methods used for the determination of residues in French beans, cauliflower, kohlrabi and cottonseed were suitable for determining the oxygen analogue separately and no residue was detectable. The other results were obtained with colorimetric methods which measured the total residue. The low level of the oxygen analogue in the crops examined indicates that the oxygen analogue does not contribute significantly to the total residue.

The meeting concluded that metabolites should be excluded from the definition of the residue. The MRLs should refer to the parent compound alone. The new definition does not alter the limits recommended previously.

4.3 AZOCYCLOTIN*

Residue and analytical aspects

To be in accord with the report of the 1982 meeting on azocyclotin/cyhexatin (para. 4.3) and noting the change in the estimated maximum residue level for cyhexatin in beans proposed by the present meeting, it was necessary to amend the proposed limit for azocyclotin to 0.2 mg/kg to maintain the comparability between residues resulting from uses of azocyclotin and cyhexatin. The meeting considered that a re-evaluation of the residues of azocyclotin on fruiting vegetables would not be worthwhile until further residues data and information on use patterns were available. It is important to recognize that good agricultural practice in the use of azocyclotin and cyhexatin is not expected to lead to the use of these pesticides together on the same crop. Consequently,the MRLs of these compounds should not be regarded as additive.

4.4 BENOMYL*

Residue and analytical aspects

New information was received on approved uses of benomyl in 22 countries. Data were available on benomyl residues in crops following seasonal treatments according to good agricultural practice in the USA, Kenya, The Federal Republic of Germany, England, Australia and Japan. The residues in commodities for which carbendazim guideline levels had been recorded supported those levels,with exceptions to be noted later,and could be used to propose additional limits for asparagus, chestnuts, pineapples, taro and sweet potatoes.

Information on the fate of benomyl in soil continues to show that the primary degradation product is carbendazim together with a small amount of 2-aminobenzimidazole (2-AB). Benomyl is strongly bound to soil and does not enter the water in flooded rice fields to any significant extent. Photodegradation on surfaces (fruit or glass) is slow, with the gradual formation of carbendazim over many weeks.

In feeding studies with a lactating goat given 5 daily doses (by capsule) of ^{14}C-labelled benomyl, equivalent to 36 mg/kg in the total daily diet, 96% of the fed benomyl was excreted in the urine and faeces.

Cooking (boiling) green beans fortified at 2 mg/kg with radiolabelled benomyl extracted slightly more than half of the residue into water (as carbendazim) with the remainder, also as carbendazim, in the beans. Because of the degree of extractability into hot water thus demonstrated, no recommendation for an MRL for tea could be proposed, even though some information was available from a supervised field trial.

An analytical survey of soybean-based products,such as infant formulas, vegetable oils etc,did not show any measurable residues (<0.05 mg/kg).

The improved method of residue analysis of Pease and Holt remains the method of choice for most commodities for regulatory use, although newer HPLC-based methods are available,which show promise for future development.

A market-basket type survey of eighteen food items from retail sources in ten US cities showed that 76% had no detectable residues. The highest residue found was 0.73 mg/kg in dried apricots.

Information on national tolerances was available to the meeting from a number of countries.

The 1978 JMPR agreed to replace GLs for benomyl by those for carbendazim. Those figures were converted to MRLs at the present meeting, on the basis of the estimation of an ADI (see below). Additional maximum residue levels for pineapples of 20 mg/kg, asparagus and taro of 0.1 mg/kg, chestnuts of 0.2 mg/kg, and sweet potato of 1 mg/kg were estimated for carbendazim on the basis of supervised trials with benomyl. The data also supported increases in the MRLs for carbendazim in wheat straw from 2 to 5 mg/kg, in melons from 0.5 to 2 mg/kg and in peanut hulls from 0.2 to 1 mg/kg. The figures were derived by dividing tabulated benomyl residues data by the conversion factor of 1.52.

Toxicology

Available information demonstrates the similarity of metabolism of benomyl and carbendazim. Benomyl is rapidly metabolized to carbendazim in mammals and is eliminated preferentially in urine as methyl 5-hydroxy-2-benzimidazolecarbamate (5-hydroxycarbendazim, 5-HBC). In rats,following intubation, inhalation and dietary exposures, benomyl and carbendazim were present in the blood within the first 6 hours, together with comparable levels of 5-HBC. Within 18 hours after exposure only 5-HBC was identified in blood. Metabolism proceeds via hydroxylation and ester hydrolysis in the liver followed by elimination, primarily in the urine (41-71%) and to a lesser extent in faeces (21-46%). There is no retention of ^{14}C activity in muscle tissue or fat, only liver and kidney demonstrating bound residues identified as 5-HBC.

Benomyl is not acutely toxic to mammals, with an acute oral LD_{50} in rats and an acute dermal LD_{50} in rabbits above 10 000 mg/kg. Gross and histopathological examination of many of these animals demonstrated a measurable compound-related effect on the male gonads at high doses (testicular degeneration, necrosis of germinal epithelium and aspermatogenesis).

Spermatogenesis in pre-pubertal rats was not affected at dose levels which affected adult rats. Previous JMPR evaluations have identified variations in teratogenic response dependant upon the mode of oral administration (diet or gavage). The present meeting reviewed additional data which confirmed the difference in sensitivity according to the route of administration in the rat, the dietary NOEL for teratogenicity being 6 760 ppm. This dose did, however, induce embryotoxicity. In the mouse, gavage studies resulted in terata induction at levels above 50 mg/kg. In the rabbit, a limited study did not result in induction of terata at 500 ppm in the diet.

Short- and long-term dietary studies in rats did not demonstrate compound-related effects at doses up to and including 2500 ppm. Ninety-day and two-year dietary studies in dogs demonstrated adverse effects on the liver at 2500 ppm, but not at 500 ppm, evidenced by increased cholesterol, alkaline phosphatase and GPT levels, by decreased total protein and albumin/globulin ratio, and by bile duct proliferation and haemosiderosis (see also 4.8 carbendazim).

A carcinogenicity study in CD-1 mice at 500, 1500 and 5000 ppm in the diet showed oncogenicity at all levels. Neoplastic changes included lung carcinomas in males, but not females, at 500 and 5000 ppm. There was increased incidence of hepatocellular carcinomas in males at 500 and 1500 ppm, and in females at 500 and 5000 ppm. There was no measurable effect on latency period. It was concluded that benomyl was hepatocarcinogenic to mice (see also carbendazim).

Mutagenicity studies with benomyl gave both positive and negative results. Benomyl was positive in the micronucleus, yeast, fungi and drosophila tests. Conflicting negative and positive results in other tests prevented evaluation of the mutagenic potential. The potential impact of these results on human health cannot be adequately assessed at this time.

The monographs on benomyl and carbendazim have stated that the metabolism of the two compounds is essentially the same, with benomyl converted rapidly to carbendazim in mammals. Accordingly, the available data for benomyl and carbendazim should be considered collectively for the evaluation of specific studies, such as teratology, reproduction, chronic toxicity and oncogenicity, taking into account the different molecular weights of the two compounds.

Previous meetings have discussed the etiology and pathogenesis of liver tumours in certain strains of mice, with particular emphasis on organochlorine pesticides (JMPR 1970, 1973, 1976). It was recognized that liver tumours are known to develop spontaneously in many strains of mice, at relatively high incidence and without intentional exposure to chemicals. Evidence of such tumours in several strains of mice has been found in many of the oncogenicity studies performed with benomyl and carbendazim. Furthermore, one strain of mouse used (HOE NMR) is known to have a low background incidence of liver tumours (1-2%) and did not provide evidence of oncogenicity when exposed to carbendazim at doses up to and including 5000 ppm. Two additional studies have been carried out in rats using both benomyl and carbendazim. Both studies were negative for oncogenicity at doses up to and including 2500 and 10 000 ppm respectively. The hepatic tumours produced in mice, therefore, appear to be a species-related phenomenon.

The meeting expressed concern at the equivocal nature of the results of a wide range of mutagenicity studies. The possibility that conflicting results were due to variations in the type and amount of impurities was considered. However, the meeting received information that current levels of the relevant impurities are very low in technical materials.

In view of established no-observed-effect levels determined in several studies, including teratology, reproduction and chronic feeding, an ADI for both benomyl and carbendazim could be estimated. However, a safety factor of 200 was used to reflect the meeting's concern for the paucity of individual animal data for many studies on carbendazim, which reflects the toxicity of benomyl.

Level causing no toxicological effect

Rat: 2500 ppm in the diet, equivalent to 125 mg/kg bw

Dog: (See Carbendazim)

Rat: 30 mg/kg bw/day (teratology)

<u>Estimate of acceptable daily intake for man</u>

0-0.02 mg/kg bw

<u>Further work or information</u>

Desirable

1. Data on individual animals in studies on carbendazim that have been identified in the monograph.

2. Additional data to elucidate the mechanism of degenerative testicular effects on mammals.

3. Elucidation of the variability of the mutagenicity data.

4.5 BITERTANOL
all-rac-1-(biphenyl-4-yloxy)-3,3-dimethyl-1-(1*H*-1,2,4-triazol-1-yl)=butan-2-ol

Bitertanol is a fungicide with protective, curative and eradicative effects on a range of diseases. Formulated as a 25% wettable powder it is recommended, for example, against scab on apples and pears.

Residue and analytical aspects

No data were available for review at the meeting.

Toxicology

After oral administration, the rat excreted radiolabelled bitertanol and its metabolites, mostly in the faeces. After 7 days, tissues contained low levels of radioactivity. Hydroxylated metabolites were identified in the faeces but details of the degree of absorption and the metabolic fate of these compounds in the rat were not elucidated.

Bitertanol is of low acute toxicity to the mouse, rat, rabbit, sheep and dog. The symptoms of acute intoxication are consistent with central nervous system toxicity. Gastrointestinal irritation occurs at high doses after oral administration.

Subchronic oral administration of bitertanol to dogs induced significant dermal lesions including inflammation, scale formation and hair loss, as well as conjunctival and gingival irritation at doses above 1 mg/kg/day.

In chronic studies, the rat was more susceptible to bitertanol toxicity than the mouse. There was no evidence of carcinogenicity in either species at doses tested. A battery of short-term mutagenicity investigations did not indicate a mutagenic potential.

Bitertanol produced maternal toxicity and some evidence of teratogenicity in rats receiving more than 10 mg/kg/day and rabbits more than 30 mg/kg/day. Clear separation of maternal toxicity and teratogenicity could not be achieved. Maternal toxicity and embryoyotoxicity but not teratogenicity were observed in the rat multigeneration reproduction study at dietary levels above 20 ppm.

Level causing no toxicological effect

Mouse: 20 ppm in the diet, equal to 3 mg/kg bw

Rat: 20 ppm in the diet, equivalent to 1 mg/kg bw

Rabbit: 30 mg/kg bw/day (based on reproduction)

Dog: 1 mg/kg bw/day

Estimate of temporary acceptable daily intake for man

0-0.005 mg/kg bw

Further work or information

Required (by 1987)

1. Metabolism studies to clarify the metabolic pathway of bitertanol in mammals.

2. Oral toxicity study in the dog with a minimum duration of one year.

3. Chronic toxicity and carcinogenicity studies in rats at appropriate dosage.

Desirable

Observations in man.

4.6 BROMIDE ION*

Residue and analytical aspects

Residues of bromide ion in food commodities and in diets were last evaluated by the 1981 JMPR. The occurrence of bromide in crops grown in soil treated with methyl bromide is recognized as a potential major contribution to bromide in the diet. A report on bromide ion residues in UK crops following soil fumigation with methyl bromide was made available to the meeting. Extensive data from supervised trials and from surveys of commercial crops were reviewed. Bromide ion levels in first crops after fumigation often exceeded the proposed Codex limit (at Step 6) of 100 mg/kg for lettuce and cabbage. Residue data reviewed enabled the meeting to estimate maximum residue levels of 50 mg/kg bromide ion in cucumbers, 300 mg/kg in celery and 75 mg/kg in tomatoes. These levels were considered suitable for proposals as MRLs. The meeting noted that in a UK programme of surveillance of retail food samples for bromide ion residues, with the exception of lettuce, celery, radishes, tomatoes and eggplant, the mean levels were all less than 10 mg/kg.

4.7 BUTOCARBOXIM
3-(methylthio)butanone *O*-methylcarbamoyloxime

Butocarboxim is a systemic insecticide of the oximecarbamate family, which is effective in foliar treatment for the control of sucking insets on cotton, fruits and vegetables. It is marketed as either a 25% or 50% a.i. emulsifiable concentrate and recommended use rates range from 0.3 to 3.5 kg/ha on most commodities, with typical pre-harvest intervals varying from 3 to 21 days. There are no post-harvest, animal or soil-treatment uses (except on ornamental plants).

Residue and analytical aspects

Butocarboxim is degraded rapidly in plants to form the sulphoxide and, mainly, sulphone metabolites. Residues of the parent compound are rarely detectable 14 days after treatment. The results of supervised trials were available for a number of fruits and vegetables, which enabled some guideline levels to be recorded. Although the second major use is on cotton (15-20% of production), the results of supervised trials in progress were not available in time for the meeting.

The method of analysis used in obtaining most of the reviewed data is based on partition of an extract into toluene and water fractions, followed by saponification of each fraction to liberate methylamine, which is then measured as the N-methyl derivative of 2,4-dinitrofluorobenzene by gas chromatography with electron-capture detection. Limits of determination of 0.02-0.04 mg/kg are usually achieved for most commodities.

An improved method was developed in 1976 in which the toluene fraction is cleaned up and the butocarboxim determined directly by gas chromatography with a nitrogen-selective thermionic detector. The metabolites in the aqueous phase are subjected to ester interchange in alkaline methanol to form methyl N-methylcarbamate, which is also determined by nitrogen-selective gas chromatography. This method is recommended for regulatory use.

Toxicology

No data were available for review at the meeting.

Further work or information

Required (by 1984)

Results of supervised trials in progress on residues in cottonseed, and in cottonseed oil resulting from the foliar treatment of cotton, in order to estimate maximum residue levels.

Desirable

1. Data on residues in meat and milk resulting from the feeding of cottonseed, cottonseed oil or citrus by-products to animals.

2. Additional residue data from supervised trials on greenhouse-grown vegetables, particularly lettuce and tomatoes.

4.8 CARBENDAZIM*

Residue and analytical aspects

New information was received on use patterns and registered uses for carbendazim in 29 countries. Some additional information was available on the fate of residues in plants and soil.

Systemic uptake of labelled carbendazim from treated soil (2 kg/ha) was slight, reaching maximum levels of total ^{14}C of 0.3 mg/kg in alfalfa, 0.53 mg/kg in soybean plants, and 0.19 mg/kg in ryegrass. Each plant contained approximately equal amounts of carbendazim, 2-aminobenzimidazole (2-AB) and a polar unknown fraction.

The mobility of carbendazim in soil is limited, with a maximum of 0.4% of applied ^{14}C being found in run-off water from artificial rain and 0.2% in leach water. The soil contained 91% of applied activity in the treated area and 93% in the first 10 cm.

Degradation of carbendazim in soil under anaerobic conditions resulted in only a small amount (<2% of the applied dose) of 2-AB and no other products. The radioactivity in the applied carbendazim was apparently mostly incorporated into soil humus.

New or additional data on residue levels in various crops resulting from supervised trials with carbendazim and benomyl were received from a number of countries. The data supported previously estimated maximum levels for a number of crops but indicated a need for increased estimates in several others: bean fodder (formerly bean vines) from 30 to 50 mg/kg, apricots, grapes and sugar-beet tops from 5 to 10 mg/kg, rice straw from 2 to 15 mg/kg, peanut hay from 2 to 5 mg/kg, and peanut hulls from 0.2 to 1. Additional commodities for which estimates could be made included dry hops (50 mg/kg), peppers and blueberries (5 mg/kg), onions (2 mg/kg), eggplant and oats (0.5 mg/kg), chestnuts and soybeans (0.2 mg/kg) and asparagus, rutabagas and soybean hay (0.1 mg/kg).

The improved residue analytical method of Pease and Holt is still recommended for regulatory use. A simple HPLC method,which individually determines residues of benomyl and carbendazim on apple leaves without clean-up,has been published, which appears highly useful for research purposes.

The guideline levels recorded in 1978 are now converted to maximum residue limits on the basis of the establishment of an ADI for carbendazim.

Toxicology

Carbendazim follows a similar metabolic pathway to benomyl in rats and mice, being excreted in urine as 5-hydroxy carbendazim (5-HBC). Enzyme induction studies demonstrate that the rat is more efficient than the mouse in metabolizing and eliminating carbendazim and its metabolites.

Carbendazim is not acutely toxic to mammals as demonstrated by acute oral and dermal LD_{50}s of >10 000 mg/kg in rat and rabbit, respectively. Gross and histopathological examinations performed in many of these acute studies indicated that doses $\geq$1 000 mg/kg produced adverse effects on the testes (small, soft, discoloured, degenerative changes of the tubules) and epididymides (reduced or absent sperm).

A three-generation reproduction study in rats demonstrated a NOEL of 500 ppm, higher doses resulting in reduced average litter weights.

Teratology studies in which the test material was administered in the diet of rats indicated in one study the absence of induction of terata at 10 000 ppm, and in the second study a low incidence of misshapen, fused or incompletely ossified bones at 6000 ppm. A limited study in rabbits did not indicate the induction of terata following dietary administration at 6000 ppm (see also 4.4 - benomyl).

A short-term dietary study in rats indicated increased liver to body weight ratios in females at 2500 ppm, although no compound-related histomorphic changes were evident. There were no effects on testicular weight and the NOEL was 500 ppm. Two short-term dietary studies in dogs demonstrated that 300 and 500 ppm, respectively,caused no adverse effects. However, at doses of 1000 and 2500 ppm, animals lost their appetite, lost weight and had increased cholesterol levels and relative liver weight increases.

In two separate long-term feeding studies in rats, carbendazim produced relative liver weight increases, deposition of pigment in the spleen and bone marrow and decreased haemoglobin, hematocrit and RBC counts at the higher doses. It was without adverse effects at 300 and 500 ppm,respectively, and there was no oncogenic response at doses up to 10 000 ppm.

Beagle dogs appeared to be more sensitive than rats to dietary exposure to carbendazim. Hepatic cirrhosis, vacuolated hepatic cells and increased levels of cholesterol, BUN, total protein, GPT and alkaline phosphatase, with decreased A/G ratio,were evidence of liver toxicity at levels greater than 100 ppm for 2 years.

Oncogenicity studies were performed using three strains of mice (CD-1, Swiss SPF and HOE-NMR.) In CD-1 mice there was a significant increase in hepatocellular carcinomas at 1500 and 7500 ppm in females, and 1500 ppm in males. However, there were also substantial histomorphic changes in the livers of male and female control groups. There was no oncogenic response at 500 ppm. Swiss mice, exposed for 80 weeks to 150, 300 and 1000-5000 ppm carbendazim,showed an oncogenic response at 5000 ppm, evidenced by significant increases in the incidence of neoplastic nodules and hepatoblastomas. There were no compound-related effects in this study at 300 ppm. HOE-NMR mice exposed to 50-5000 ppm carbendazim for 96 weeks presented no evidence of oncogenic response. It was concluded that carbendazim was hepatocarcinogenic to mice at high dose levels.

Mutagenicity studies with carbendazim gave both positive and negative results. Carbendazim was positive in the micronucleus, yeast, fungi and drosophila tests. Conflicting negative and positive results in other tests prevented evaluation of the mutagenic potential. The potential impact of these results on human health cannot be adequately assessed at this time.

The data for benomyl and carbendazim have indicated that the metabolism of the two compounds is essentially the same, with benomyl converted rapidly to carbendazim in mammals. Accordingly, the available data for benomyl and carbendazim should be considered collectively for the evaluation of specific studies,such as teratology, reproduction, chronic toxicity and oncogenicity, taking into account the different molecular weights of the two compounds.

Previous meetings have considered the aetiology and pathogenesis of liver tumours in certain strains of mice, with particular emphasis on organochlorine pesticides (JMPR 1970, 1973, 1976). It was recognized that liver tumours are known to develop spontaneously in many strains of mice, at relatively high incidence and without intentional exposure to chemicals. Evidence of such tumours in several strains of mice has been found in many of the oncogenicity studies performed with benomyl and carbendazim. Furthermore, one strain of mouse used (HOE-NMR) is known to have a low background incidence of liver tumours (1-2%) and did not provide evidence for oncogenicity when exposed to carbendazim at doses up to and including 5000 ppm. Two additional studies have been carried out in rats using both benomyl and carbendazim. Both studies were negative for oncogenicity at doses up to and including 2500 and 10 000 ppm, respectively. The hepatic tumours produced in mice, therefore, appear to be a species-related phenomenon.

The meeting expressed concern at the equivocal nature of the results of a wide range of mutagenicity studies. The possibility that conflicting results were due to variations in the type and amount of impurities was considered, but the meeting was informed that current levels of the impurities in question are very low in technical material.

In view of established no-observed-effect levels determined in several studies, including teratology, reproduction and chronic feeding, an ADI for both benomyl and carbendazim could be estimated. However, a safety factor of 200 was used to reflect the concern of the meeting for the paucity of individual animal data for many studies on carbendazim.

Level causing no toxicological effect

Rat: 500 ppm in the diet, equivalent to 25 mg/kg bw

Dog: 100 ppm in the diet, equivalent to 2.5 mg/kg bw

Rat: Teratology (See benomyl)

Estimate of acceptable daily intake for man

0-0.01 mg/kg/bw

Further work or information

Desirable

1. Data on individual animals used in studies on carbendazim that have been identified in the monograph.

2. Additional data to elucidate the mechanism of degenerative testicular effects on mammals.

3. Elucidation of the variability of the mutagenicity data.

4.9 CARBOPHENOTHION*

Residue and analytical aspects

The CCPR asked the JMPR to reconsider the definition of the residue. The results of supervised trials on beans, rape, cotton and citrus fruit indicate that the oxygen analogue of the parent compound and its sulphoxide and sulphone are either present in very low concentration or not detectable. Consequently, they can be excluded from the definition of the residue. The proportion of parent compound in the total residue was also considered, but most of the results obtained in supervised trials were presented as a sum including sulphoxide and sulphone. An estimation of maximum residue levels based on the parent compound alone was therefore not possible. The meeting concluded that the oxygen analogue of the parent compound and its sulphoxide and sulphone could be excluded from the definition of the residue. The MRLs, which remain numerically the same, refer to the sum of carbophenothion, its sulphoxide and its sulphone, expressed as carbophenothion.

4.10 CHINOMETHIONAT*

Residue and analytical aspects

Chinomethionat was last evaluated by the 1981 JMPR when information on use patterns on tomatoes and additional residues data from supervised trials were considered desirable. Limited new data together with those available to the 1981 meeting were reviewed, but the meeting was again unable to estimate a maximum residue level for tomatoes owing to deficiencies in the data. The meeting reaffirmed that further data were desirable.

4.11 CHLOROTHALONIL*

Residue and analytical aspects

The meeting reviewed additional information on GAP for grapes, residue trials data for grapes, onions and strawberries and monitoring data for several commodities. Data were insufficient to estimate a maximum residue level for strawberries, but were consistent with the current limits for cauliflower, celery, cucumber, melons and onions. The data on grapes suggested that the current temporary limit may be too low, but were not conclusive. The limit should remain temporary pending the receipt of sufficient data to support a permanent limit, irrespective of the status of the ADI.

Information provided on the effects of cooking on residues of chlorothalonil in treated food did not fully answer the questions previously raised, but the meeting concluded that the matter had been adequately considered. A study of surface and maceration extraction techniques for chlorothalonil and its metabolites in oranges showed comparable results by the two methods.

In keeping with Section 3.2 of this report, the meeting reviewed the definition of the residue in order to simplify regulatory analysis. Noting that the major metabolite contributes a maximum of 10% of the total residue in commodities which has been analysed, and usually much less, the meeting concluded that maximum residue limits could be expressed as chlorothalonil only, without any need to change the numerical values. The commodity description for "lima beans" was revised to reflect the data on which the MRL was based.

Toxicology

Excretion and elimination studies in rats demonstrated that chlorothalonil is preferentially excreted in the faeces, with minor elimination of the parent or its metabolites in the urine. 4-hydroxy-2,3,5-trichloroisophthalonitrile was identified as a major metabolite. Faecal elimination was complete (95%) within 24 hours on low doses, but was noticeably delayed (96 hours) at high doses. The total amount of radioactivity in the blood following the oral administration of labelled chlorothalonil was less than one percent of the administered dose, with most of the activity in the plasma. This information failed to clarify the general metabolism of chlorothalonil, particularly regarding the 4-hydroxy metabolite.

Short-term dietary studies in rats and mice exposed to chlorothalonil demonstrated increased incidences of hyperplasia and hyperkeratosis in the non-glandular portion of the stomach and hyperplasia of the epithelial cells of the proximal convoluted tubules in the kidney. Kidney and liver weights were also increased, but there was no microscopic evidence of morphological change. The NOEL was 3 mg/kg bw in both rats and mice.

A short-term dietary study in rats, using the 4-hydroxy metabolite at levels of 10-750 mg/kg, presented evidence of anaemia and compound-related liver effects at dietary levels greater than 20 mg/kg bw.

A one-generation study in rats using the 4-hydroxy metabolite demonstrated adverse reproductive effects at doses greater than 30 ppm. These included poor pup viability and reduced mean live pup weights.

Teratology studies with rabbits and rats gavaged with chlorothalonil produced no teratogenic responses; however, maternal toxicity was evident at $\geq$5 mg/kg and 400 mg/kg in rabbits and rats, respectively. Teratogenicity studies by gavage in the rabbit at maternally toxic doses of 5 mg 4-hydroxy metabolite/kg failed to induce terata. A wide range of mutagenicity studies using both chlorothalonil and the 4-hydroxy metabolite did not indicate mutagenic activity for either compound.

A carcinogenicity study in mice exposed to chlorothalonil at dietary levels of 150-3 000 ppm produced glandular epithelial tumours in treated groups, which were compound-related but not dose-related. The neoplastic changes in the kidney, including adenomas and carcinomas of the renal cortical tubules, were increased in males in all treated groups but not in females. There was no positive trend for these kidney adenomas and carcinomas and therefore the evidence for the tumourigenicity of chlorothalonil in the kidney remains elusive. A no-observed-effect level was not determined.

Carcinogenicity studies in rats and mice using the 4-hydroxy metabolite did not demonstrate oncogenic potential at doses up to and including 1 500 ppm in the diet (22.5 mg/kg bw) and 20 mg/kg bw in mice and rats, respectively. There was, however, clear evidence of anaemia at doses greater than 3 mg/kg bw in rats. A no-observed-effect level for non-oncogenic effects was not demonstrated in the mouse study.

The meeting expressed concern over the lack of adequate metabolism data for chlorothalonil and the 4-hydroxy metabolite and the need to clarify the incidence of kidney adenomas and carcinomas in rodents. The meeting was informed that a long-term oncogenicity feeding study with chlorothalonil in the rat is currently conducted, but the results would not be available until 1985. The meeting, therefore, agreed that the temporary ADI should remain at the present level until further data are made available to the JMPR to clarify remaining questions on metabolism and carcinogenicity of chlorothalonil.

Level causing no toxicological effect

Rat: 10 ppm in the diet, equivalent to 0.5 mg/kg bw

Dog: 120 ppm in the diet, equivalent to 3 mg/kg bw

Estimate of temporary acceptable daily intake for man

0-0.005 mg/kg bw

Further work or information

Required (by 1985)

1. Further work to elucidate the metabolism of chlorothalonil and the 4-hydroxy metabolite.

2. Submission of the on-going long-term oncogenicity feeding study in the rat.

3. Additional and more specific information on good agricultural practice in the use of chlorothalonil on grapes as well as residue data from field trials in countries where uses on grapes are known to be good agricultural practice.

4. Good agricultural practice information, if it exists, for those countries in which residue trials for data provided to the 1983 JMPR were conducted.

Desirable

1. Information on data that may clarify the type of anaemia which has been demonstrated in mammals exposed to the 4-hydroxy metabolite.

2. Analyses of chlorothalonil-treated animal feed items (for example, bean and peanut vines), processed and unprocessed, for residues of PCBN (from the 1981 JMPR).

3. Information on possible PCBN residues in tissues and milk of dairy cattle fed a diet containing chlorothalonil (from the 1981 JMPR).

4.12 CHLORPYRIFOS*

Residue and analytical aspects

The meeting considered data from trials in which chlorpyrifos was applied pre-harvest to grapes to provide protection to mature, drying and dried vine fruit. Residues data on the dried fruit enabled the meeting to estimate a maximum residue level for dried vine fruit of 2 mg/kg. This was considered suitable for the establishment of an MRL. The meeting confirmed that the residue should be defined as the parent compound only.

4.13 COUMAPHOS*

Residue and analytical aspects

The CCPR asked the JMPR to reconsider the definition of the residue. No data were available for evaluation which showed separately the concentration of the parent compound and its oxygen analogue except some in the 1978 evaluation on cattle milk, in which the maximum concentration of the parent compound was 0.27 and 0.16 mg/kg at 6 and 30 hours after the last application while the oxygen analogue was not detectable (<0.1 mg/kg). Because of the lack of suitable residues data a change in the definition of the residue could not be recommended.

4.14 CYANOFENPHOS*

Toxicology

Data reviewed by the 1980 JMPR included several studies carried out by Industrial Bio-Test Laboratories (IBT), which are known to be invalid. Studies published since 1980 did not alleviate the meeting's concern, particularly regarding delayed neurotoxic potential of cyanofenphos. As the meeting was advised that the additional data required by the 1980 JMPR could not be provided and the manufacture of the compound had ceased, the existing temporary ADI was withdrawn.

4.15 CYHEXATIN*

Residue and analytical aspects

Cyhexatin is the preferred miticide for the control of phytophagus mites on Kiwi fruit in New Zealand. Residue data from five trials over three seasons provided a basis for estimating a maximum residue level. Levels in the edible part of the fruit were up to about 15% of the whole fruit residue. Since maximum residue limits are recommended for Kiwi fruit on a whole-commodity basis, 5 mg/kg was proposed on the basis of a 3 to 4 day pre-harvest interval. The meeting reviewed the limited data on residues of cyhexatin on beans in the 1978 evaluations and concluded that a maximum residue level of 0.2 mg/kg was more appropriate than the figure of 0.5 mg/kg previously estimated.

Additional residues data on peaches from Canada and the USA showed that an increase in the estimate of the maximum residue level to 5 mg/kg was necessary to cover good agricultural practices in these countries.

4.16 CYPERMETHRIN*

Residue and analytical aspects

At the request of the CCPR the meeting re-examined the data supporting the estimate of 5 mg/kg (on a dry weight basis) as a maximum residue level on alfalfa, maize fodder and sorghum fodder. Taking into account the range of initial deposits of cypermethrin following treatment, the uncertainty of the period between treatment and use as a feed and the fact that some of the data referred to the green crop, the meeting found no basis for changing its earlier estimate for these animal feeds.

4.17 DAMINOZIDE*

Toxicology

Daminozide was reviewed by the 1977 JMPR but an ADI was not estimated. The rat teratogenicity study reviewed at that time was carried out by IBT and is invalid. A study subsequently published is inadequate for the evaluation of teratogenicity. Consequently,the meeting deferred further consideration of this compound.

Further work or information

Required (before an ADI can be estimated)

1. Adequate data to assess the carcinogenicity of daminozide.
2. An adequate teratogenicity study.
3. Information on biotransformation in animals.

4.18 DDT*

Toxicology

The meeting, in examining the conditional ADI on DDT, noted that the present estimate was based on a 23-week rat study utilizing 8 animals/sex/dose. Further, it was noted that ongoing studies were due for review to present a re-assessment of the conditional ADI in 1971. This review and evaluation were not carried out. The present meeting was of the opinion that presently known levels of DDT residues are unlikely to be hazardous to human health. Nevertheless, the meeting was reluctant to maintain the conditional ADI, in view of the questionable basis of the data, which does not meet current requirements for the establishment of an ADI. The meeting was aware that extensive data are available on DDT and its metabolites, which have never been reviewed by the JMPR.

Therefore, since the conditional ADI has been accepted by the JMPR since 1969, and since the meeting believed that a full data review (which is overdue) could lead to the estimation of an ADI, it reluctantly agreed to maintain the present conditional ADI. This agreement was reached only on the assumption that a review of an adequate toxicology data base is carried out prior to the next meeting and that the information in this review will be placed before the 1984 JMPR for complete evaluation.

The meeting recommended that the need for an immediate and comprehensive review of the toxicology of DDT be brought to the urgent attention of the Directors-General of FAO and WHO as well as those of other interested organizations. In the absence of such a review by the 1984 JMPR the meeting proposed that the withdrawal of the existing conditional ADI should again be considered.

4.19 DEMETON AND RELATED COMPOUNDS*

Toxicology

Demeton and related compounds were evaluated by the 1982 JMPR, which recommended withdrawal of the ADI. Subsequent to this recommendation, additional data were submitted for consideration to the 1983 JMPR. These data were received too late to be fully evaluated and are recommended for review by the 1984 JMPR.

4.20 DICHLOFLUANID*

Residue and analytical aspects

Information was received on approved uses from Finland (on apples, strawberries and currants with a waiting period of 14 days), The Netherlands (on grapes, eggplants and courgettes) and Norway (on stone fruits, berries, lettuce, onions, tomatoes and cucumbers).

Surveys of food in commerce received from Belgium, Finland and The Netherlands indicated that dichlofluanid was rarely found, never exceeding the national MRL of 5 mg/kg in Belgium and exceeding the MRL of 5 mg/kg in about 3-4% of the samples in The Netherlands. In Finland, residues exceeded 1 mg/kg in only 15 food samples of 1 081 analysed. Supervised trials at recommended rates on greenhouse tomatoes in Norway gave a maximum residue of 0.7 mg/kg, 2 days after a fifth bi-weekly treatment.

Research data from the manufacturer on hydrolysis and photostability indicated that decomposition in water was rapid, that phodegradation on a TLC plate resulted in 30-42% remaining after 4 weeks under artificial light and that dichlofluanid had a photolytic half-life of 10 hours in aqueous solution.

Since an ADI has now been estimated, all previous temporary MRLs are replaced by MRLs at the same levels.

Toxicology

The 1979 meeting, which extended the temporary ADI for dichlofluanid, required studies to elucidate the accumulation seen in the thyroid. A study on rats, over a period of 9 weeks, investigated the dose/time relationship of the thyroid effect at dose levels up to 4500 ppm. Increased iodine uptake was observed early in the study at 4500 ppm, while decreases in T3 and T4 levels were noted on days 7 and 21 of the study. The NOEL for the thyroid effects was greater than 1500 ppm. However, decreased body weight gain indicates an overall NOEL of 500 ppm for the study.

Additional data evaluated included a sister-chromatid exchange study, which was negative at doses of 2000 mg/kg, and a teratogenicity study in the rabbit, which demonstrated embryotoxicity, but not teratogenicity, at a dose level of 100 mg/kg bw. A mouse long-term carcinogenicity study did not indicate any evidence of oncogenicity at dose levels up to 1000 ppm in the diet.

Since all the requirements of the 1979 meeting have been satisfied the present meeting was able to estimate an ADI.

Level causing no toxicological effect

Rat: 500 ppm in the diet, equivalent to 30 mg/kg/bw

Dog: 1000 ppm in the diet, equivalent to 25 mg/kg/bw

Estimate of acceptable daily intake for man

0-0.3 mg/kg/bw

Further work or information

Desirable

Observations in man.

4.21 DIFLUBENZURON*

Residue and analytical aspects

The meeting reviewed the data on residues in mushrooms in the 1981 evaluations and concluded that a maximum residue level of 0.1 mg/kg was more appropriate than the figure of 0.2 mg/kg previously estimated. It was recognized, however, that at the highest rate of application studied there was a possibility that 0.1 mg/kg might be marginally exceeded at the first cropping. The meeting noted that the data supporting the 1981 estimates for maximum residue levels in Brussels sprouts, cabbage and plums had been accidentally omitted from the 1981 evaluations. An addendum correcting this error will be included in the 1983 evaluations. On reviewing these data the meeting concluded that they would not support the maximum residue level for plums estimated in 1981 and recommended that the MRL for plums be withdrawn until further data became available.

4.22 DIMETHOATE*

Residue and analytical aspects

The 1981 meeting discussed the relationship between dimethoate, omethoate and formothion and made a series of recommendations aimed at harmonizing the MRLs for these pesticides. It agreed to review the compounds at a future meeting with the aim of establishing separate MRLs for dimethoate and omethoate. When this matter was considered at the 14th and 15th Sessions of the CCPR,it was recommended that all three compounds should be reviewed together in the light of comments from governments and other interested bodies. The meeting was also aware that dimethoate had been reviewed in the USA and The Netherlands and considered that the outcome of these reviews should be taken into consideration.

It was therefore agreed that the proposed review of dimethoate should be postponed until 1984 and that it should also include omethoate and formothion, with a view to determining whether there is a systematic relationship between the concentrations of the respective residues when any of the compounds is used.

4.23 DITHIOCARBAMATE FUNGICIDES*

Residues and analytical aspects

The 1980 JMPR reviewed the dithiocarbamates and required several items of further work or information by 1983, including further development and comparison of analytical procedures, information on good agricultural practice and residue data for thiram, ferbam, ziram and propineb, and residue data for dithiocarbamates on crops grown under glass. Further studies on other ethylenebisdithiocarbamate (EBDC) fungicides,which are ethylenethiourea (ETU) precursors,were considered desirable. Limits for dimethyldithiocarbamates, ethylenebisdithiocarbamates and propylenebisdithiocarbamates were considered temporary until the requirements were met.

Except for propineb, the only information on dithiocarbamates for review at this meeting was on national tolerances for dithiocarbamates, good agricultural practices for thiram (in Spanish) and a submission on EBDCs (in French). (See para 2.9 .) For propineb, the meeting received information on GAP, residue trials data for a limited number of fruits and vegetables from one country and plant metabolism data. However, all dithiocarbamate submissions were received too late to permit evaluation and will be considered by the 1984 JMPR. All dithiocarbamate maximum residue levels are still considered temporary until the submitted data are evaluated and other outstanding requirements are met.

4.24 ETHEPHON*

Residue and analytical aspects

Ethephon was evaluated by the JMPR in 1977 and 1978. As no ADI was estimated,only guideline levels were proposed. New information was available to the meeting on uses of ethephon on cereals and new data were received from supervised trials on a number of vegetables, fruit and cereals. As the residue data from trials on cereals originated from a country with no registered uses of ethephon, no maximum residue levels could be estimated for cereal grains.

4.25 ETHIOFENCARB*

Residue and analytical aspects

At the request of the CCPR the meeting reviewed the limited data on ethiofencarb in the 1977 monographs on beans, both with and without pod. The meeting agreed that for beans with pod the earlier data together with the data in the 1981 evaluations could be interpreted as supporting a maximum residue level of 2 mg/kg but could find no basis for changing its earlier estimate of 0.2 mg/kg for beans without pod.

The relevant data are re-presented in the 1983 monograph on ethiofencarb.

4.26 ETHION*

Residue and analytical aspects

The CCPR asked the JMPR to reconsider the definition of the residue. The original data showed that ethion residues are oxidized to the mono- and di- oxygen analogues,which are subsequently hydrolysed. The rate of degradation on or in plants decreases in the order di-oxon>mono-oxon>ethion. No di-oxon derivative was detected in any plant samples except maize forage, where its concentration was approximately 5-10% of the total residue. The proportions of the mono-oxon derivative in the total residue are summarized in the 1972 evaluations, from which it can be seen that the total residue in dried prunes and raisins consists of approximately 25% and 50% mono-oxon,respectively. The meeting concluded that the metabolites could be excluded from the definition of the residue and that the MRLs should refer to the parent compound only. The new definition does not alter the numerical values of the limits recommended previously.

4.27 ETHOPROPHOS
O-ethyl *S*,*S*-dipropyl phosphorodithioate

Ethoprophos is a non-systemic non-fumigant nematicide and soil insecticide. It is an organophosphorus compound, formulated as a 10% granule and a 70% emulsifiable concentrate; rates of use are from 1.6-6.6 kg ai/ha for most crops.

Residue and analytical aspects

No data were available for review at the meeting.

Toxicology

A limited excretion and elimination study in the rat identified *O*-ethyl *S*-propyl phosphorothioate as a major metabolite in the urine.

Ethoprophos is acutely toxic to rats, mice and rabbits, with an oral LD_{50} ranging from 30 to 60 mg/kg body weight.

A 3-generation reproduction study in rats demonstrated no adverse reproductive effects at a dietary level of 60.5 ppm. Teratology studies in rats and rabbits were negative for teratogenic effects at 0.16 mg/kg and 2 mg/kg bw, respectively. The mutagenicity studies available, including unscheduled DNA synthesis in primary rat hepatocytes, a mouse lymphoma assay and an *in vivo* cytogenetic study in rats, were negative. A 109-week oncogenicity study in rats showed an increased incidence of thyroid C-cell adenomas in surviving high-dose males. Compound-related cholinesterase inhibition was evident in all dose groups for both plasma and brain enzymes. RBC cholinesterase was not affected. A NOEL was not demonstrated in this study.

A delayed neurotoxicity study in hens was inadequate.

A short-term study in rats did not demonstrate a NOEL, since cholinesterase levels were depressed in all treatment groups. A 90-day dog study demonstrated no adverse effects on cholinesterase at 1 ppm.

The meeting could not estimate an ADI because the data base was inadequate.

4.28 FENCHLORPHOS*

Residue and analytical aspects

The CCPR asked the JMPR to reconsider the definition of the residue. Thin-layer chromatographic tests referred to in the 1968 evaluation indicate that the concentration of the oxygen analogue in samples of animal origin is negligible. In the experience of members of the CCPR Ad Hoc Working Group on Methods of Analysis, the oxygen analogue was not found in samples moving in commerce. The exclusion of the oxygen analogue from the definition of the residue is justified and the meeting concluded that MRLs should refer to the parent compound only. The new definition does not alter the numerical values of the MRLs recommended previously.

4.29 FENITROTHION*

Residue and analytical aspects

A maximum residue level for fenitrothion in oranges was estimated by the 1974 JMPR. The mode of application of the compound has been changed since then.

Results of supervised trials reflecting current good agricultural practice indicate residues in whole mandarins (satsumas) ranging from 0.2 mg/kg to 2.1 mg/kg and in oranges of about 1 mg/kg 28 days after application.

The residue was mainly concentrated in the peel and the pulp contained residues of 0.03 mg/kg or less. The residue levels found in shipments imported by Finland from Spain were consistent with the results of supervised trials.

The major routes of degradation of fenitrothion are hydrolysis, demethylation and reduction to the amino compound. The formation of fenitro-oxon is minor compared to that of products formed via hydrolysis. No fenitro-oxon was detected in the milk or meat of cattle kept on diets containing fenitrothion residues. The oxygen analogue may be formed in plants but it occurs only during the initial period at a low concentration (ca. 1% of the total residue) and its half-life is estimated to be only a few hours. Apples, rice, wheat and wheat products were analysed for metabolites of fenitrothion but no oxon was reported.

On the basis of limited information derived from supervised trials carried out in accordance with present GAP and the results of food surveys, the meeting concluded that the existing maximum residue limit (0.2 mg/kg) for fenitrothion in oranges should be increased to 2 mg/kg. The limit is temporary regardless of the status of ADI and applies also to mandarins (satsumas).

The meeting recommended that all temporary MRLs should refer to the parent compound only. The numerical values of the limits remain the same.

<u>Further work or information</u>

Required (by 1986)

1. Information on current good agricultural practices.

2. Results of supervised trials carried out on different varieties of citrus fruits according to present use recommendations.

4.30 FENSULFOTHION*

<u>Residue and analytical aspects</u>

The CCPR asked the JMPR to reconsider the definition of the residue. The original data showed that the residues in supervised trials were measured after oxidation with *p*-chloroperbenzoic acid, which converts all the residues to the sulphone of the oxygen analogue. The proportions of the residue components were not measured in any of the crops for which MRLs are greater than the limit of determination, in spite of the availability of an analytical procedure which would determine the metabolites separately. The meeting concluded that the definition of the residue could not be changed on the basis of available data.

4.31 FENTHION*

<u>Residue and analytical aspects</u>

The CCPR asked the JMPR to reconsider the definition of the residue. The original data showed that the major metabolite in plants is the sulphoxide of the parent compound. Its sulphone and the sulphoxide and sulphone of the oxygen analogue are present at much lower levels. In contrast, the sulphoxide of the oxygen analogue is the major metabolite (ca. 97%) in milk. No information was available on the proportions of the residue components in meat or in many of the plant commodities. The meeting concluded that the definition of the residue could not be changed on the basis of the available data.

4.32 MECARBAM*

Residue and analytical aspects

The meeting reviewed additional information on citrus fruit and grapes. Information on GAP for citrus fruit showed that the current pre-harvest interval was now longer than that on which earlier data were based.

The new citrus residue data supported the level of 2 mg/kg previously estimated for oranges and, together with data previously reviewed, indicated that the level should also be applicable to all citrus fruit. Swedish import monitoring data also gave added assurance that the level is adequate for oranges, grapefruit and lemons, and also demonstrated the need for an MRL at that level for this relatively persistent pesticide in citrus. The data also confirmed that residues in citrus are mostly in the peel and demonstrated that up to 30% of the terminal residue in the peel is the oxygen analogue of mecarbam.

In the absence of information on approved or registered uses of mecarbam on grapes, and in view of the relatively limited amount of residue data or information on the analytical method used, a maximum residue level for grapes could not be estimated.

Toxicology

The 1980 JMPR estimated a temporary ADI with a requirement for metabolic studies, a teratogenicity study, mutagenic studies and a delayed neurotoxicity study.

An interim report of a study with ^{14}C-labelled mecarbam indicated rapid absorption, complete metabolism and excretion mainly via the urine. Over 80% excretion of the administered dose occurred within 48 hours, the major excretory products comprising 7 polar metabolites.

There was no evidence of teratogenic activity in rats dosed at 3 mg/kg. Mutagenicity studies, including *in vitro* microbial assays, unscheduled DNA synthesis in cultured human epithelial cells and a micronucleus test in mice, were negative. However, a mouse lymphoma test, at highly toxic dose levels resulting in about 20% cell survival, seemed to elicit a positive mutagenic response. An *in vivo* cytogenic test in rats was inconclusive under the conditions of the experiment. A delayed neurotoxicity study in young adult hens was seemingly negative. However, histopathological examination of nerve tissue was confined to only 3 of the 6 surviving hens. Moreover, the dosage level used (equivalent to an oral LD_{50}) is considered insufficient for a delayed neurotoxicity study.

Since the only metabolic studies available (presented as an interim report) were carried out on the rat and the delayed neurotoxicity study was not acceptable, the meeting could only extend the temporary ADI estimated in 1980.

Level causing no toxicological effect

Rat: 5 ppm in the diet, equivalent to 0.21 mg/kg bw

Dog: 5 ppm in the diet, equivalent to 0.15 mg/kg bw

Estimate of temporary acceptable daily intake for man

0-0.001 mg/kg bw

Further work or information

Required (by 1985)

1. An adequate delayed neurotoxicity study in hens.

2. Full report of the metabolic studies in rats.

3. Complete metabolic studies in laboratory animals other than the rat.

4. A ruminant metabolism study.

5. Should the ruminant metabolism study indicate any possibility of residues of concern in animal tissues or milk, other than those identified in plants, data should be provided on these residues occurring in meat and milk from feeding ruminants with a diet containing residues found in treated citrus.

Desirable

1. Additional information on nationally registered or approved uses on additional commodities (especially olives, olive oil, fruits and vegetables) and residue field trials data reflecting those uses.

2. Observations in man.

4.33 METHIOCARB*

Residue and analytical aspects

From the information received since the original evaluation in 1981 it is clear that many countries recognize the use of methiocarb bait (4%) against slugs and snails in vegetable, grain, fruit and sugarbeet crops. Further information on the fate of methiocarb in animals, plants, soil and water and the effect of light on deposits and solutions confirmed the evaluations made in 1981. It is clear that methiocarb is rapidly degraded in plants, water and soil so that there is little possibility of carry-over in crop rotation. Because methiocarb is rapidly metabolized in animals,there is also unlikely to be any transfer of residues in to foods of animal origin following the feeding of forage or food offals to livestock. However, this applies only to residues from current practices and does not allow for residues that might arise from grazing or feeding livestock on stubbles or other parts of plants from grain crops sprayed with methiocarb to repel birds. Data from Thailand confirmed several existing recommendations and enabled the meeting to estimate a maximum residue level for Chinese radish.

It was suggested at the 15th Session of the CCPR that the limit of determination in regulatory practice should be 0.1 mg/kg, not 0.02 mg/kg as proposed by the 1981 JMPR. The meeting reconsidered this question and noted that for several commodities the apparent residue on untreated control samples was sometimes higher than that reported on the treated samples. Bearing in mind the comment on limits of determination in para 2.4 of the report of the 1981 JMPR, it seemed reasonable to expect that food control officials might have difficulty in working to a limit of determination as low as 0.02 mg/kg. The meeting agreed not to recommend MRLs lower than 0.05 mg/kg.

Some delegations to the 15th Session of the CCPR questioned whether it was good agricultural practice to use methiocarb as a bird repellent close to harvest to prevent destruction of such crops as blueberries, cherries, currants and grapes. There can be no doubt that a number of countries have accepted such uses as being essential to the protection of these crops and studies made in Belgium indicated that when methiocarb is used to protect cherries from birds the residues can be up to 15 mg/kg. It was noted that cereal grain crops also are being protected from attack by birds in Africa, New Zealand, the USA and elsewhere.

The meeting gave special consideration to the toxicological significance of the residues on apples, blueberries, cherries, currants, grapes and peaches arising from the use of methiocarb as a bird repellent. While recognizing that the majority of the fruits would normally contain methiocarb residues at concentrations considerably lower than the MRLs, the meeting considered the possibility that these commodities, consumed fresh, might contain residues that could lead to an intake of methiocarb that exceeded the demonstrated NOEL for plasma cholinesterase in dogs noted by the 1981 JMPR. The meeting did not have available for review the details of this dog experiment and therefore was unable to evaluate fully the significance of this demonstrated inhibition of plasma cholinesterase. Attention was drawn to para 3.2 of the report of the 1982 JMPR dealing with the function of plasma cholinesterase and the meeting agreed to request that the relevant information should be made available to the 1984 JMPR to allow full consideration of the safety of the observed residue levels.

The meeting recommended that the MRLs for the commodities in question should be withdrawn pending this further review.

Toxicology

The cholinesterase studies in both rats and dogs indicated methiocarb sulphoxide to be a more potent inhibitor than methiocarb. No evidence of teratogenicity of methiocarb was observed in the rabbit teratology study.

The meeting confirmed the ADI estimated at the 1981 meeting and further desirable work was recommended.

Level causing no toxicological effect

Rat: 25 ppm in the diet, equivalent to 1.3 mg/kg bw

Dog: 5 ppm in the diet, equivalent to 0.125 mg/kg bw

Estimate of acceptable daily intake for man

0-0.001 mg/kg bw

Further work or information

Desirable

1. Data to clarify the significance of the inhibition of plasma cholinesterase noted in the dog study which led to the estimation of the NOEL by the 1981 JMPR, bearing in mind the views of the 1982 JMPR on the function of plasma cholinesterase.

2. Information on the method of analysis used to estimate plasma and RBC cholinesterase activities *in vivo* in the above dog study.

3. Observations in man.

4.34 NITROFEN
2,4-dichlorophenyl 4-nitrophenyl ether

Nitrofen is a selective pre- and early post-emergence herbicide for the control of annual grasses and various broad-leaved weeds in cereal grains and vegetables. Its use is limited to three countries and recommendations for its use either prohibit women from handling the product or demand stringent handling precautions that prevent exposure during application. At present its uses are limited, mainly to wheat and brassicas, with application rates up to 2 kg ai/ha.

Residue and analytical aspects

Samples from supervised trials on broccoli, cabbage, cauliflower, celery, garlic, rape and rice contained no measurable residues (<0.01 mg/kg) regardless of dosage, pre-harvest interval or analytical method. In many of the wheat samples the parent compound was not detectable, in others the total residue measured as the amino derivative ranged up to 0.09 mg/kg. Wheat straw contained the parent compound at levels of <0.002 to 0.02 mg/kg, while the sum of metabolites and parent ranged from <0.01 mg/kg to 0.11 mg/kg.

The metabolic pathways of nitrofen include reduction of the nitro group, conjugation of the resulting amino group, hydroxylation of the dichlorophenyl ring and incorporation into naturally-occurring materials. The major metabolites identified in plants and animals are largely similar. In rice plants, the parent compound accounted for 72% of the extractable residue at day 10, while hydroxyacetamidonitrofen (15%), the parent compound (13.7%) and the amino derivative (11.8%) were the major components after 90 days. A similar pattern of metabolite formation was observed in wheat. Neither the grain nor the straw contained extractable residues at harvest. In studies with the radiolabelled compound, 64-69% of the radioactive carbon was found in the starch of rice and wheat grains. Approximately 50% of the unextractable activity was found in the lignin fraction of the rice straw.

Metabolites were identified in the urine, faeces and blood of sheep dosed with the labelled compound at the extreme rate of 40 mg/kg. In urine, the bulk of the radioactivity consisted of sulphate, glucuronide and glycine conjugates. The organosoluble fraction (2 to 5.5% of the total residue) consisted mainly of aminonitrofen and hydroxynitrofen, while the parent compound amounted to only 0.3%. In blood, the parent compound averaged 33.9% of the total activity, while aminonitrofen and unknown compounds amounted to 20% and 13.4%, respectively.

Feeding studies on cattle dosed with ring-labelled nitrofen by capsule revealed that the total radioactive residue, expressed as parent compound, in milk was 0.003 mg/kg at a dose level equivalent to 0.5 mg/kg diet, while at a rate of 5 mg/kg the residue was 0.08 mg/kg, 0.24 mg/kg and 1.15 mg/kg in milk, muscle and fat, respectively. However, no residue was detectable in the milk, urine or faeces of cattle when the nitrofen was mixed with the ration at a level of 5 mg/kg, owing to the rapid degradation of the parent compound and aminonitrofen in rumen fluid.

Eggs and tissues of laying hens kept on a diet containing 0.04 mg/kg, 0.12 mg/kg and 0.48 mg/kg radiolabelled nitrofen were analysed. After a 6-week feeding period the residue in eggs reached a plateau at levels of 0.02, 0.05 and 0.17 mg/kg, respectively. Almost all the residue in eggs was found in the yolk. Tissue residues ranged from non-detectable in white meat to 0.18 mg/kg in fat at the 0.04 mg/kg feeding level.

As the lowest feeding level is approximately 10 times the calculated maximum residue which might result from treated feed components, no detectable residue can be expected in the milk or tissues of cattle or eggs or meat of poultry.

Under field conditions the nitrofen concentration in or on soil declined moderately. The major metabolites identified in the soil were aminonitrofen and hydroxyacetamidonitrofen. Both nitrofen and its aged derivatives were immobile in soil, which is in keeping with their high adsorption coefficient on different soils and the low water solubility of the parent compound.

Analytical methods are available for the determination of the parent compound alone, or the majority of the organo-extractable residues after derivatization of aminonitrofen. The former method is recommended for regulatory purposes.

The meeting concluded that the data available were sufficient for estimating maximum residue levels in wheat grain and straw. As an ADI was not estimated, these were recorded as guideline levels. They refer to the parent compound only.

Toxicology

Nitrofen has low acute oral toxicity. Following oral administration, nitrofen is partially absorbed and rapidly excreted. A small proportion of the administered dose is distributed to all tissues and organs. Nitrofen and its metabolites are found primarily in faeces (accounting for about 75% of the administered dose), while urine contains metabolites accounting for about 15% of the dose. The metabolism of nitrofen involves reduction, hydroxylation and conjugation.

Two reproduction studies were presented. Because of inadequacies in the reports the studies were not considered.

Oral exposures to nitrofen during the period of organogenesis produced neonatal mortality, decreases of survival to weaning and a series of abnormalities in rodent offspring. In a mouse teratology/post-natal study, the lowest dose tested (6.25 mg/kg bw) produced increased incidence of lethal defects at birth and resulted in decreased lung and seminal vesicle weights at maturity. In a rat teratology study, diaphragmatic hernias were observed at and above 1.39 mg/kg bw but full details were not provided.

Nitrofen in some cases was found mutagenic in the Salmonella/microsome assay but it did not display any mutagenic potential in several mammalian systems, either *in vitro* or *in vivo*.

Two carcinogenicity studies on B6C3F1 mice indicated that nitrofen is a liver carcinogen, causing hepatocellular carcinomas and hepatocellular adenomas in both sexes and haemangiosarcomas in male mice. In addition, the compound is carcinogenic to female Osborne-Mendel rats, causing adenocarcinomas of the pancreas. Nitrofen was not found to be carcinogenic in Fischer 344 rats. Detailed histological examinations suggest that naturally occurring and nitrofen-induced liver tumours in mice are morphologically different. Both studies were of short duration. In the study on Osborne-Mendel rats, poor survival precluded the evaluation of carcinogenicity in males.

In a long-term rat feeding study a NOEL of 10 ppm was observed. However, the duration of the study (97 weeks), the poor survival in all groups (50% beyond the 65th week) and the fact that the study was not available *in extenso* precluded its full evaluation.

An ADI could not be estimated, owing to the evidence of carcinogenicity, the lack of a NOEL for the teratology and post-natal effects and the inadequacies of several studies, including reproduction and long-term studies.

<u>Further work or information</u>

Desirable

Data on nitrofen residues in wheat bran.

4.35 OXAMYL*

<u>Residue and analytical aspects</u>

Oxamyl was first evaluated in 1980. Since then, some additional data have been made available on use patterns and on residues from supervised trials. When the use recommendations were followed, the residue in whole bananas ranged from less than 0.01 mg/kg to 0.14 mg/kg, while in the pulp it was somewhat lower. Residues in coffee beans were usually undetectable. They were measurable in only a few cases, with a maximum of 0.08 mg/kg. Residues in dried coffee beans were higher than in the corresponding green ones.

No residue was detectable in mature maize. The green forage contained residues in the range 0.26 to 2.6 mg/kg and below 0.1 mg/kg 4 and 8 weeks after treatment, respectively. Soybean samples contained maximum residues of 0.05 mg/kg, while no residue was detectable in pressed oil or meal. Following soil treatment the oxamyl residue in sugarcane at harvest was below 0.02 mg/kg. The residue levels in cucumbers grown in glasshouses did not decrease significantly during the first two weeks after foliar application.

No dimethylcyanoformamide (DMCF) residue was detected in any of the samples, except green maize forage in which the maximum level found was 0.02 mg/kg.

The meeting concluded that the temporary MRLs previously proposed for bananas and soybeans should be increased,while that for cucumbers should be lowered. Maximum residues levels were estimated for coffee beans, onions and sugarcane and these were regarded as suitable for establishing temporary MRLs.

Residue data on maize, sugarbeet and tomato supported the previous recommendations but data on Brussels sprouts, French beans, gherkins, lettuce and strawberry were insufficient to estimate maximum residue levels.

Toxicology

Oxamyl was first evaluated in 1980,when a temporary ADI was estimated. Indentification of tissue residues and clarification of the no-effect level in the rat, especially in relation to the marginal effect seen at the 50 ppm dietary level, were required. Metabolism studies with oxamyl and some of its metabolites, as well as acute and short-term studies,and a mutagenicity study with dimethylcyanoformamide (DMCF), one of the metabolites in vegetables, were received too late to be evaluated fully.

While the metabolism studies may satisfy the first requirement,the toxicological studies submitted do not satisfy the other requirements of the 1980 JMPR. The meeting therefore agreed to extend the temporary ADI at the same level until 1985 and reaffirmed the necessity of clarification of the no-effect level in the rat with oxamyl *per se*.

Level causing no toxicological effect

Dog: 100 ppm in the diet,equivalent to 2.5 mg/kg bw

Estimate of temporary acceptable daily intake for man

0-0.01 mg/kg bw

Further work or information

Required (by 1984)

1. Further data on beans, for reconsideration of the temporary recommendations.

2. Additional data on materials used for animal feedstuffs, e.g. sugarbeet leaves and bean fodder.

Required (by 1985)

3. Clarification of the no-effect level in the rat, especially in relation to the marginal effects of 50 ppm on the body weight in several studies.

Desirable

1. Additional residue data from trials reflecting GAP in additional countries; in particular, residue data on peas.

2. Information on residues in foods in commerce and at sonsumption.

3. Information on the effect on oxamyl residues of cooking, processing or storage of raw agricultural commodities.

4. Crop rotation studies on additional commodities, under field conditions, with applications of commercial formulations, both granular and liquid, according to maximum recommendations.

5. Observations in man.

4.36 PERMETHRIN*

Residue and analytical aspects

Additional residues data enabled the meeting to estimate maximum residue levels for sorghum fodder and peanuts. It was considered that the level for peanuts should apply to the group of legume oilseeds. New residues data on phaseolus beans indicated that the existing MRL of 0.5 mg/kg could be exceeded in good agricultural practice and the meeting estimated an increased maximum level of 1 mg/kg. The commodity description should be changed from beans (whole green) to beans (with pod). All three estimates were considered suitable for establishing MRLs.

4.37 2-PHENYLPHENOL AND ITS SODIUM SALT*

Toxicology

The meeting considered new information on both sodium 2-phenylphenate (sodium *o*-phenylphenate, SOPP) and 2-phenylphenol (*o*-phenylphenol, OPP) and agreed that these compounds should be considered as equivalent for purposes of toxicological evaluation and should be dealt with together. The use of SOPP results in OPP residues in agricultural commodities.

SOPP has been established as a bladder carcinogen in the rat; fewer data on OPP are available but a similar pattern has been reported. Accordingly, the meeting agreed to convert the existing ADI for 2-phenylphenol and its sodium salt to a temporary ADI and to increase the safety factor to reflect the concerns of the meeting. In addition, the meeting agreed to seek the additional information which would permit adequate evaluation in the future.

Level causing no toxicological effect

Dog: 500 mg/kg bw/day

Rat: 2000 ppm in the diet, equivalent to 100 mg/kg bw

Estimate of temporary acceptable daily intake for man

0-0.02 mg/kg bw

Further work or information

Required (by 1985)

1. Information on the progress of a multigeneration reproduction study.

2. Information on the progress of a carcinogenicity/chronic toxicity study in a rat strain known to be sensitive to induction of bladder carcinomas (a strain other than Fischer F 344).

3. Metabolic, pharmacokinetic and other related studies, as appropriate, in the strains of animals tested in long-term studies for carcinogenicity.

4. Additional metabolic and pharmacokinetic studies in other species and strains of animals.

5. Information on current use patterns and on the levels and fate of 2-phenylphenol resulting from current good agricultural practice.

4.38 PHORATE*

Toxicology

The 1982 JMPR evaluated phorate and estimated a temporary ADI, requiring the submission of an appropriate delayed neurotoxicity study by 1983. The meeting was informed that the required data were not available. However, in the light of information provided by WHO and the manufacturer, the meeting agreed to extend the temporary ADI until 1985.

Level causing no toxicological effect

Rat: 1 ppm in the diet, equivalent to 0.05 mg/kg bw

Dog: 1 ppm in the diet, equivalent to 0.01 mg/kg bw

Estimate of temporary acceptable daily intake for man

0-0.0002 mg/kg bw

Further work or information

Required (by 1985)

An appropriate delayed neurotoxicity study.

4.39 PHOXIM*

Residue and analytical aspects

The toxicology of this compound was evaluated in 1982. The meeting then was unable to estimate maximum residue levels as there were no clear indications of the uses in good agricultural practice. The meeting has now received information on practical uses, which together with information received earlier on metabolism in plants and soil and data from supervised trials, enabled the meeting to estimate maximum residue levels for cereals, several vegetables and the meat of cattle and sheep.

Phoxim is used on vegetables both as a soil- and foliage-applied insecticide. Soil applications are made at planting or sowing, mainly on potatoes. Foliage application is mainly used on cotton and maize for the control of lepidopterous larvae. Residues are usually small and, in most cases, below the limit of determination.

Phoxim is also used for animal treatments against ectoparasites and mites. Residues in liver, kidney and muscle from the recommended treatment of sheep and cattle were below the limit of determination, but residues in the fat from the same animals were up to about 0.5 mg/kg.

The metabolites in plants consist of the thiolo-isomer, together with tetraethyl pyrophosphate and tetraethyl monothiopyrophosphate. Phosphorus-free metabolites, formed by hydrolysis, react with plant constituents to form three glycosides.

Metabolism has been studied in rats and mice, but no information has yet been received on metabolism in sheep, cattle or pigs after treatment with phoxim.

Phoxim is degraded in soil with a half-life from 1 day to about 2 weeks, although the half-life in a greenhouse study was about 8 weeks.

Analytical methods suitable for regulatory purposes are available. Residues are determined by GLC with a flame-photometric or thermionic detector.

4.40 PIRIMIPHOS-METHYL*

Residue and analytical aspects

Pirimiphos-methyl was evaluated in 1974, 1976, 1977 and 1979 and a number of items of information previously requested are still outstanding.

It would appear that there are a number of uses that give rise to residues in food moving in international trade which have not yet been considered by the meeting and about which relevant information will be available in 1984.

Extensive information about the use and fate of pirimiphos-methyl when applied post-harvest to cereal grain has appeared in open scientific literature and this was reviewed by the meeting. The review confirms the evaluation made previously. All authors have drawn attention to the stability of pirimiphos-methyl deposits on treated grain and the retention of the bulk of the deposit in the seed coat. This is reflected in the low transfer of residue into white flour and milled rice and the higher residues in wheat bran.

This review has confirmed that the degradation and metabolism of pirimiphos-methyl on grain does not lead to the production of cholinesterase-inhibiting compounds. However, between 4% and 15% of the residue of the parent compound forms a complex with the aleurone layer of the grain and resists extraction by a variety of solvents. It may be dislodged by digesting the substrate with proteolytic enzymes.

A further study of the use of pirimiphos-methyl for the control of cheese mites in cheese stores confirmed that the MRL previously recommended adequately covers the resulting residues.

Several reviews have been made of analytical methods suitable for dealing with various substrates.

The information evaluated by the meeting indicated that there was no need to amend or modify recommendations previously made for MRLs in a range of commodities. It was concluded that the metabolites could be excluded from the definition of the residue. The MRLs, which remain the same, refer to the parent compound alone.

The meeting reviewed the requests made by earlier meetings for additional information in the light of the known current uses of this insecticide and considered that they had been satisfied.

Further work or information

Desirable

Further information about residues in peanuts, oil seeds, lentils and citrus.

4.41 PROCHLORAZ
N-propyl-*N*-[2-(2,4,6-trichlorophenoxy)ethyl]imidazole-1-carboxamide

Prochloraz is used for the control of a variety of fungal diseases on a number of crops. It is available as 40% and 45% emulsifiable concentrates for foliar and post-harvest treatments, a 50% wettable powder containing a 4:1 prochloraz-manganese co-ordination complex and other formulations for specific uses.

Residue and analytical aspects

The meeting reviewed residue data from a number of countries from field trials of prochloraz on a variety of commodities. Most were foliar applications to cereal grains, fruit and a few vegetables. Data from post-harvest treatments of citrus and some other fruit and from uses during the culture of mushrooms were also reviewed. Most of the data were for total residues of prochloraz and its metabolites; prochloraz was analysed separately in some cases. Unfortunately information on nationally approved or registered uses was available only for watermelons, apples, grapes, cereals, mushrooms and oil seed rape. In order to facilitate enforcement (see Section 2.3) maximum residue limits were recommended only for commodities for which there were data specifically on prochloraz. On those for which only "recommended" uses were available and those for which residues were determined as the sum of prochloraz and its metabolites, only temporary maximum residue limits are recommended and further data are required.

Animal metabolism data demonstrate fairly rapid urinary or faecal excretion of metabolic products. Urinary metabolites have been characterized, but those found in tissues have not, although the relative distribution in tissues has been studied. Residues are typically highest in the liver.

The major plant residues are the "formyl urea" and the "urea", with lesser amounts of 2,4,6-trichlorophenol and 2,4,6-trichlorophenoxyacetic acid and with conjugation increasing with time from application. Residues translocate only to a small extent from soil or foliar applications and are typically located largely in the peel of fruit after foliar treatments. Residues are fairly persistent under a variety storage conditions on the commodities tested, especially on citrus.

Analytical methods are available for the determination of prochloraz alone or the sum of prochloraz and its metabolites.

Toxicology

Prochloraz has low oral toxicity. Following oral administration in the rat, 95% of the dose is rapidly absorbed. In rats, the major excretion is via the urine; in dogs it is via the faeces. Prochloraz is a demonstrated liver mixed-function oxidase inducer.

In rats and rabbits, no evidence of teratogenic effects was noted at maternally toxic dose levels. A multigeneration reproduction study in rats did not show any adverse effects at doses up to 37.5 ppm. Mutagenicity studies, including the Ames test, micronucleus test, mouse lymphoma assay and a dominant lethal study,were all negative.

In 90-day studies in mice, rats and dogs, liver weight was increased in all species. No-effect levels of 6 mg/kg/day and of 2.27 mg/kg/day were established for mice and dogs, respectively. However,in rats 6 mg/kg/day induced increased liver weight and occasional signs of intoxication (increased salivation, diarrhoea). A two-year dog study established a NOEL of 30 ppm.

A long-term rat study did not result in any evidence of oncogenic potential. A NOEL of 37.5 ppm in the diet was established. A mouse carcinogenicity study of 106 weeks duration indicated an increased incidence of liver adenomas and adenocarcinomas in both sexes at dose levels of 325 ppm and above. Prochloraz has been determined to be a hepatocarcinogen to mice.

Level causing no toxicological effect

Rat (male): 37.5 ppm in the diet, equal to 1.3 mg/kg bw

Dog: 30 ppm in the diet, equal to 0.9 mg/kg bw

Estimate of acceptable daily intake for man

0-0.01 mg/kg bw

Further work or information

Required (by 1985)

1. In order to establish MRLs for products of animal origin:

   a. metabolism studies in animals, including ruminants and poultry, at sufficiently high dosing levels to permit identification and quantification of tissue residues.

   b. if metabolism studies in (a) indicate any possibility of residues in animal tissues, feeding trials should be conducted in animals, including ruminants and poultry, in which tissues are analysed for residues of concern identified in the metabolism studies. Feeding in an appropriate number of animals should as nearly as possible reflect maximum levels of weathered residues likely to occur in practice, and ideally at an exaggerated level as well.

   c. a validated analytical method suitable for regulatory purposes for products of animal origin.

2. Good agricultural practice information for those commodities for which temporary maximum residue limits are recommended, preferably for countries in which the residue trials were conducted or those in close proximity.

3. Additional residue trials data reflecting good agricultural practice for commodities for which temporary maximum residue limits are recommended (except citrus). Both prochloraz and total residues of prochloraz and metabolites containing the trichlorophenol moiety should be determined on the same samples. Data should be sufficient, together with residue data already provided, to enable maximum residue limits, expressed as prochloraz alone, to be recommended.

Desirable

1. Additional national tolerances.

2. Additional processing studies in cereals to permit a conclusion on the extent, if any, of residue concentration in milling fractions. Residues in the grain used should be high enough for valid conclusions to be drawn.

3. Additional processing studies for mushrooms to permit a conclusion on the extent, if any, of residue concentration. Mushrooms should be analysed before and after processing.

4. Information on possible interference in prochloraz analysis by other pesticides, especially those containing the trichlorophenol moiety. If interferences are encountered, confirmatory analytical procedures for prochloraz and its metabolites may be necessary.

5. Observations in man.

4.42 PROPINEB*

Toxicology

The meeting examined the available data on propineb and concluded that a complete evaluation was not possible. In addition, since only histograms of data, and incomplete reports, were available on long-term studies on propylenethiourea (PTU), the meeting deferred consideration of this compound to the next JMPR. Meanwhile, the temporary ADI was extended for two years.

Level causing no toxicological effect

Rat: 10 ppm in the diet, equivalent to 0.5 mg/kg bw

Dog: 3000 ppm in the diet, equivalent to 75 mg/kg bw

Estimate of temporary acceptable intake for man

0-0.005 mg/kg bw

Further work or information

Required (by 1985)

Submission of adequate data for evaluation.

4.43 PROPOXUR*

Residue and analytical aspects

The CCPR at its 14th Session asked the JMPR to reconsider the definition of the residue.

Propoxur disappears from treated surfaces mainly by volatilization and only small amounts penetrate through the surfaces of plant leaves. Since the surface residue is not, or is only slightly, affected by photo-decomposition, the major proportion of it is the parent compound. The portion of the surface residue which penetrates into the plant is rapidly converted to the water-soluble metabolites 2-hydroxyphenyl *N*-methylcarbamate (metabolite A) and 2-isopropoxyphenyl *N*-hydroxymethylcarbamate (metabolite B). The metabolites are conjugated with plant glucosides and are not extractable with organic solvents. In bean plants the parent compound was 50.1% and 58.7% of the organo-soluble residues after one and three days, respectively.

The results of supervised trials on fruit and vegetables listed in the 1973 evaluations were obtained by a method which measured the parent compound only. The residues in field crops were determined by a specific method and conjugated metabolites were released by enzyme hydrolysis before clean-up. The proportion of parent compound and metabolites varied after application. The parent compound accounted for about 20-50% of the total residue in cereal grains, green cereal forage, grass and alfalfa. The cereal straw contained the parent compound only.

The meeting concluded that the metabolites could be excluded from the definition of the residue. The MRLs should refer to the parent compound alone. As a result of the new definition and the reconsideration of experimental results, the limits for cereal grains (0.5 mg/kg), fodders and straws (5 mg/kg) and legume animal feeds (green) (5 mg/kg) need to be reduced. MRLs for fruits and vegetables should remain the same.

4.44 TECNAZENE*

Toxicology

The 1981 JMPR re-evaluated tecnazene and at that time required that an adequate multigeneration reproduction study should be made available for evaluation by the present meeting. No such study has been provided.

Previous meetings considering this compound have evaluated two multigeneration studies, both of which were deemed to be inadequate. The first study reported data only on fertility and gross pathology. No effects were observed at dose levels exceeding 1000 ppm. The second study was more comprehensive but was performed on inadequate numbers of animals. Again, however, no positive effects were identified at dose levels exceeding 2000 ppm. Furthermore, tecnazene did not show embryotoxic or teratogenic properties when administered orally at doses up to 200 mg/kg to one strain of rat and two strains of mouse, from day 7 to day 16 or 18 of gestation.

Comparative toxicological information on tecnazene and quintozene (which are closely related compounds) indicate that their metabolic patterns are essentially similar. Results of long-term studies, teratology studies,etc are also comparable, the liver being the primary target organ. No teratogenic effects have been observed at dose levels up to 200 mg/kg/day.

In the light of the above facts,the meeting considered it reasonable to utilize the data on an adequate reproduction study on quintozene to support the two inadequate studies on tecnazene reviewed by previous meetings. Consequently,the meeting estimated a full ADI for tecnazene but retained the safety factor previously utilized for the temporary ADI.

Level causing no toxicological effect

Rat: 750 ppm in the diet,equivalent to 38 mg/kg bw

Mouse: 1500 ppm in the diet,equivalent to 200 mg/kg bw

Dog: 15 mg/kg bw/day

Estimate of acceptable daily intake for man

0-0.01 mg/kg bw

Further work or information

Desirable

Studies on metabolism in different mammalian species.

4.45 THIRAM*

Toxicology

Thiram was reviewed by the 1980 JMPR and additional data on teratogenic potential and studies to resolve the effects of thiram,including anaemia, were required by 1983. These data have not yet been submitted and the meeting agreed to extend the temporary ADI for two years after which consideration should be given to the withdrawal of the temporary ADI.

Further work or information

Required (by 1985)

The information required by the 1980 meeting, specified in para 4.14 items 5 and 6.

4.46 TRIADIMEFON*

Residue and analytical aspects

Triadimefon was evaluated in 1979 and 1981. New residue data based on the determination of the parent compound and the main metabolite, "triadimenol", have been received from trials on fruit, beets, coffee and cereals and from feeding experiments on cattle and hens. Most of the data confirmed the residue limits proposed in 1979, but the meeting was also able to estimate some new maximum residue levels.

The need for increased residue limits for grains and straw had been suggested, but the data received supported only an increase in the limit for wheat grain.

Studies on the uptake, distribution and metabolism in plants confirmed previous studies showing that the main metabolite is triadimenol, but that glycoside conjugates of triadimenol and 4-chlorophenol are also present in plants treated with triadimefon.

Toxicology

Studies in rats and rabbits confirmed previous findings that triadimefon has little or no teratogenic potential. The signs of acute poisoning indicate central nervous system toxicity, but investigations on the mechanism of this were inconclusive. *In vitro*, triadimefon inhibited hepatic microsomal de-ethylation of 7-ethoxycoumarin, aldrin epoxidation and biphenyl hydroxylation, and bound to cytochrome P-450. In rats and mice fed with triadimefon, induction of 7-ethoxycoumarin de-ethylation and aldrin epoxidation, but not of biphenyl hydroxylation or cytochrome P-450 levels, occurred. These results confirm previous findings but do not explain either the treatment-related increase in liver weights or the hyperplastic liver nodules previously reported in rats and mice. The meeting agreed to extend the temporary ADI until 1985.

Level causing no toxicological effect

Mouse: 300 ppm in the diet, equivalent to 40 mg/kg bw

Rat: 50 ppm in the diet, equivalent to 2.5 mg/kg bw

Dog: 330 ppm in the diet, equivalent to 8.25 mg/kg bw

Estimate of temporary acceptable daily intake for man

0-0.01 mg/kg bw

Further work or information

Required (by 1985)

Clarification of the toxicological significance of hepatic toxicity in rats and mice and of the hyperplastic liver nodules observed in chronic mouse feeding studies.

Desirable

Observations in man.

4.47 TRIAZOPHOS*

Residue and analytical aspects

Triazophos was evaluated toxicologically in 1982, when a temporary ADI was estimated. The uses, residues resulting from such uses and the fate of the residues were not considered at that time.

Triazophos is registered in more than 30 countries and the meeting received information on use patterns in 18 countries where MRLs and withholding periods have been established. The application rate ranges from 40 g to 2 kg/ha.

The meeting received reports of 318 residue trials together with corresponding analytical results. These indicated that, owing to the relatively low rate of application, the initial deposits are generally less than 2 mg/kg. The residual half-life under rain-free conditions ranges from 2 to 4 weeks,depending on the crop, with a considerable reduction following rain. This is because the major proportion of the deposit remains on the surface. Though triazophos penetrates plant surfaces,it is not translocated to any measurable extent. The fully esterified residues at harvest consist almost exclusively of the parent compound, the oxygen analogue being either absent or occurring at extremely low concentrations.

No information was available on the fate of triazophos residues in livestock but,from studies carried out on rats,it appears likely that triazophos and its degradation products are rapidly excreted in the urine and faeces and that little or none is retained in body tissues or fat.

Triazophos has been shown not to be leached from soil, nor to affect biological processes in soil, and to be degraded at a moderately rapid rate in soils typical of several parts of the world. It is not taken up from soil by plants,although it may be absorbed from hydroponic solutions.

Information was available on the effect of processing and refining on residues in cottonseed and cottonseed oil: the triazophos residues in the cottonseed oil were destroyed when the oil was hydrogenated. A study on Chinese cabbage showed that 50% of the triazophos residue was removed by water washing and 50% of the remainder was destroyed by cooking.

Triazophos residues may be extracted from plant materials with acetone and determined after transfer to methylene chloride by gas chromatography,using a flame-photometric detector. The method gives good recoveries over the range 0.01-1.0 mg/kg,with a limit of determination of 0.005 mg/kg.

The meeting estimated a number of maximum residue levels,which were regarded as suitable for establishing temporary MRLs. These refer to triazophos; metabolites are not included.

Further work or information

Desirable

Information on the fate of triazophos in livestock.

5. RECOMMENDATIONS

5.1

The meeting expressed great concern over the continuing delay in the proposed review of the toxicology of DDT for the purpose of estimating an acceptable daily intake for man. It recommended that the matter be brought to the urgent attention of the Directors-General of FAO and WHO as well as those of other interested organizations, indicating that the current conditional ADI was based on data that do not meet current scientific standards.

5.2

The meeting drew attention to para 3.2 in the report of the 1982 JMPR and recognized that uncertainty in evaluating the significance of the depression of plasma cholinesterase at low feeding levels in dog studies had caused problems to the present meeting in considering certain uses of methiocarb (para 4.33). It recommended that the problem be studied as a matter of some urgency at the 1984 JMPR.

6. FUTURE WORK

The following items should be considered in 1984 or at a future meeting.

1. Pesticides scheduled at the present or at a previous meeting for re-evaluation in 1984. These include:

   from 1980 meeting - amitraz, imazalil, maleic hydrazide, phenthoate;

   from 1981 meeting - *sec*-butylamine, chinomethionat, diflubenzuron;

   from 1982 meeting - acephate, bendiocarb, captan, captafol, chlordane, diphenylamine, fenitrothion, fenvalerate, folpet, parathion-methyl, phenothrin, phoxim;

   from 1983 meeting - oxamyl (FAO).

2. Pesticides recommended for priority attention by the 15th Session (1983) of CCPR. These are:

   cyhalothrin, dimethipin, flucythrinate, propamocarb, carbosulfan, methoprene.

3. Pesticides listed for evaluation in 1983 but not considered fully for various reasons. These are:

   bitertanol, butocarboxim, demeton and related compounds, ethoprophos, dithiocarbamate fungicides, fenarimol, isoprocarb, terbufos, thiram.

4. Pesticides for which data have become available for evaluation, namely dimethoate and omethoate.

5. DDT,for which a specific recommendation has been made (see para 5.1).

FURTHER WORK OR INFORMATION FOR FUTURE MEETINGS

The present report lists 'Requirements' for a number of compounds for review in a specific year. These are summarized below:

By 1984 butocarboxim (FAO), oxamyl (FAO).

By 1985 chlorothalonil (FAO, WHO), mecarbam (FAO, WHO), methiocarb (WHO), oxamyl (WHO), 2-phenylphenol (FAO, WHO), phorate (WHO), prochloraz (FAO), propineb (WHO), thiram (WHO), triadimefon (WHO).

By 1986 fenitrothion (FAO).

By 1987 bitertanol (WHO).

7. PREVIOUS FAO AND WHO DOCUMENTATION

FAO/WHO. 1958 Procedures for the testing of international food additives to establish their safety for use. Second Report of the Joint FAO/WHO Expert Committee on Food Additives. FAO Nutrition Meetings Report Series, No.17: WHO Technical Report Series, No.144.

FAO/WHO. 1961 Evaluation of the carcinogenic hazards of food additives. Fifth Report of the Joint FAO/WHO Expert Committee on Food Additives. FAO Nutrition Meetings Report Series, No. 29: WHO Technical Report Series, No.220.

FAO/WHO. 1962 Principles governing consumer safety in relation to pesticide residues. Report of a meeting of a WHO Expert Committee on Pesticide Residues held jointly with the FAO Panel of Experts on the Use of Pesticides in Agriculture. FAO Plant Production and Protection Division Report, No.PL:1961/11: WHO Technical Report Series, No.240.

FAO/WHO. 1964 Evaluation of the toxicity of pesticide residues in food; report of a Joint Meeting of the FAO Committee on Pesticides in Agriculture and the WHO Expert Committee on Pesticide Residues. FAO Meeting Report, No.PL:1963/13: WHO/Food Add./23 (1964).

FAO/WHO. 1965a Evaluation of the toxicity of pesticide residues in food. Report of the Second Joint Meeting of the FAO Committee on Pesticides in Agriculture and the WHO Expert Committee on Pesticide Residues. FAO Meeting Report, No.PL:1965/10: WHO Food Add./26.65.

FAO/WHO. 1965b Evaluation of the toxicity of pesticide residues in food. FAO Meeting Report, No.PL:1965/10/1: WHO Food Add./27.65

FAO/WHO. 1965c Evaluation of the hazards to consumers resulting from the use of fumigants in the protection of food. FAO Meeting Report, No.PL:1965/10/2: WHO Food Add./28.65.

FAO/WHO. 1967a Pesticide residues in food. Joint report of the FAO Working Party on Pesticide Residues and the WHO Expert Committee on Pesticide Residues. FAO Agricultural Studies, No. 73: WHO Technical Report Series, No. 370.

FAO/WHO. 1967b — Evaluation of some pesticide residues in food. FAO:PL/CP/15: WHO Food Add./67.32.

WHO. 1967 — Procedures for investigating intentional and unintentional food additives. Report of a WHO Scientific Group. WHO Technical Report Series, No. 348.

FAO/WHO. 1968a — Pesticide residues. Report of the 1967 Joint Meeting of the FAO Working Party and the WHO Expert Committee. FAO Meeting Report, No.PL:1967/M311: WHO Technical Report Series, No. 391.

FAO/WHO. 1968b — 1967 evaluation of some pesticide residues in food. FAO PL:1967/M/11: WHO Food Add./68.30.

FAO/WHO. 1969a — Pesticide residues in food. Report of the 1968 Joint Meeting of the FAO Working Party of Experts on Pesticide Residues and the WHO Expert Committee on Pesticide Residues. FAO Agricultural Studies, No. 78: WHO Technical Report Series, No. 417.

FAO/WHO. 1969b — 1968 evaluation of some pesticide residues in food. FAO PL:1968/M/9/1: WHO Food Add./69.35.

FAO/WHO. 1970a — Pesticide residues in food. Report of the 1969 Joint Meeting of the FAO Working Party of Experts on Pesticide Residues and the WHO Expert Group on Pesticide Residues. FAO Agricultural Studies, No. 84: WHO Technical Report Series, No. 458.

FAO/WHO. 1970b — 1969 evaluations of some pesticide residues in food. FAO PL:1969/M/17/1: WHO Food Add./70.38.

FAO/WHO. 1971a — Pesticide residues in food. Report of the 1970 Joint Meeting of the FAO Working Party of Experts on Pesticide Residues and the WHO Expert Group on Pesticide Residues. FAO Agricultural Studies, No. 87: WHO Technical Report Series, No. 474.

FAO/WHO. 1971b — 1970 evaluations of some pesticide residues in food. AGP 1979/M/312/1: WHO Food Add./71.42.

FAO/WHO. 1972a — Pesticide residues in food. Report of the 1971 Joint Meeting of the FAO Working Party of Experts on Pesticide Residues and the WHO Expert Committee on Pesticide Residues. FAO Agricultural Studies, No. 88: WHO Technical Report Series, No. 502.

FAO/WHO. 1972b — 1971 evaluations of some pesticide residues in food. AGP:1971/M/9/1: WHO Pesticide Residues Series, No. 1.

FAO/WHO. 1973a — Pesticide residues in food. Report of the 1972 Joint Meeting of the FAO Working Party of Experts on Pesticide Residues and the WHO Expert Committee on Pesticide Residues. FAO Agricultural Studies, No. 90: WHO Technical Report Series, No. 525.

FAO/WHO. 1973b — 1972 evaluations of some pesticide residues in food. AGP:1972/M/9/1: WHO Pesticide Residues Series, No. 2.

FAO/WHO. 1974a — Pesticide residues in food. Report of the 1973 Joint Meeting of the FAO Working Party of Experts on Pesticide Residues and the WHO Expert Committee on Pesticide Residues. FAO Agricultural Studies, No. 93: WHO Technical Report Series, No. 545.

FAO/WHO. 1974b — 1973 evaluations of some pesticide residues in food. AGP: 1973/M/9/1: WHO Pesticide Residues Series, No. 3.

WHO. 1974 — Assessment of the carcinogenicity and mutagenicity of chemicals. Report of a WHO Scientific Group. WHO Technical Report Series, No. 546.

FAO/WHO. 1975a — Pesticide residues in food. Report of the 1974 Joint Meeting of the FAO Working Party of Experts on Pesticide Residues and the WHO Expert Committee on Pesticide Residues. FAO Agricultural Studies, No. 97: WHO Technical Report Series, No. 574.

FAO/WHO. 1975b — 1974 evaluations of some pesticide residues in food. AGP: 1974/M/11: WHO Pesticide Residues Series, No. 4.

FAO/WHO. 1976a — Pesticide residues in food. Report of the 1975 Joint Meeting of the FAO Working Party of Experts on Pesticide Residues and the WHO Expert Committee on Pesticide Residues. FAO Plant Production and Protection Series, No. 1: WHO Technical Report Series, No. 592.

FAO/WHO. 1976b — 1975 evaluations of some pesticide residues in food. AGP: 1975/M/13: WHO Pesticide Residues Series, No. 5.

FAO/WHO. 1977a — Pesticide residues in food. Report of the 1976 Joint Meeting of the FAO Panel of Experts on Pesticide Residues and the Environment and the WHO Expert Group on Pesticide Residues. FAO Food and Nutrition Series, No. 9. FAO Plant Production and Protection Series, No. 8: WHO Technical Report Series, No. 612.

FAO/WHO. 1977b — 1976 evaluations of some pesticide residues in food. AGP: 1976/M/14.

FAO/WHO. 1978a — Pesticide residues in food. Report of the 1977 Joint Meeting of the FAO Panel of Experts on Pesticide Residues and the Environment and the WHO Expert Group on Pesticide Residues. FAO Plant Production and Protection Paper 10 Rev.

FAO/WHO. 1978b — 1977 evaluations of some pesticide residues in food. FAO Plant Production and Protection Paper 10 Sup.

FAO/WHO. 1978c — Index of Documentation and Summary Recommendations as at January 1978. FAO AGP:1978/M/5.

FAO/WHO. 1979a — Pesticide residues in food. Report of the 1978 Joint Meeting of the FAO Panel of Experts on Pesticide Residues in Food and the Environment and the WHO Expert Group on Pesticide Residues. FAO Plant Production and Protection Paper 15.

FAO/WHO. 1979b — 1978 evaluations of some pesticide residues in food. FAO Plant Production and Protection Paper 15 Sup.

FAO/WHO. 1980a. Pesticide residues in food. Report of the 1979 Joint Meeting of the FAO Panel of Experts on Pesticide Residues in Food and the Environment and the WHO Expert Group on Pesticide Residues. FAO Plant Production and Protection Paper 20.

FAO/WHO. 1980b. 1979 evaluations of some pesticide residues in food. FAO Plant Production and Protection Paper 20 Sup.

FAO/WHO. 1981a. Pesticide residues in food. Report of the 1980 Joint Meeting of the FAO Panel of Experts on Pesticide Residues in Food and the Environment and the WHO Expert Group on Pesticide Residues. FAO Plant Production and Protection Paper 26.

FAO/WHO. 1981b. 1980 evaluations of some pesticide residues in food. FAO Plant Production and Protection Paper 26 Sup.

FAO/WHO. 1982a. Pesticide residues in food. Report of the 1981 Joint Meeting of the FAO Panel of Experts on Pesticide Residues in Food and the Environment and the WHO Expert Group on Pesticide Residues. FAO Plant Production and Protection Paper 37.

FAO/WHO. 1982b. 1981 evaluations of some pesticide residues in food. FAO Plant Production and Protection Paper 37 Sup.

FAO/WHO. 1983a. Pesticide residues in food. Report of the 1982 Joint Meeting of the FAO Panel of Experts on Pesticide Residues in Food and the Environment and the WHO Expert Group on Pesticide Residues. FAO Plant Production and Protection Paper 46.

FAO/WHO. 1983b. 1982 evaluations of some pesticide residues in food. FAO Plant Production and Protection Paper 49.

ANNEX I

ACCEPTABLE DAILY INTAKES, MAXIMUM RESIDUE LIMITS AND GUIDELINE LEVELS PROPOSED AT THE 1983 MEETING

These figures are additional to, or amend, those recorded in Annexes of the reports of earlier meetings. Limits recommended at meetings from 1965-1977 inclusive are summarized in document FAO/WHO 1978c.

This table includes maximum acceptable daily intakes (ADIs), maximum residue limits (MRLs) and guideline levels (GLs). These terms are defined in Annex 3 of the Report of the 1975 Meeting. Part I lists ADIs and MRLs. Some ADIs and MRLs are temporary; in these cases the year in which further data are required is specified in parentheses below the ADI. Part 2 lists GLs.

New, as distinct from amended, recommendations are identified by the symbol * before the commodity or before the ADI. Residue levels at or about the limit of determination are followed by the symbol **.

Amended recommendations are followed by the previous recommendation in parentheses.

PART 1. - ACCEPTABLE DAILY INTAKES (ADIs) and MAXIMUM RESIDUE LIMITS (MRLs)

Pesticide and year of previous evaluations	ADI (mg/kg body weight)	Commodity	Recommended MRL (mg/kg)	Remarks
amitraz 1980	0.0005 (1984)	carcase meat of sheep	0.1 (0.2)	The previously recommended temporary MRL of 0.5 mg/kg in cucumbers is reaffirmed. MRLs are for the sum of amitraz, calculated as *N*-(2,4-dimethylphenyl)-*N*-methylformamidine, and *N*-(2,4-dimethylphenyl)-*N*-methylformamidine.
azocyclotin 1979, 1981, 1982	0.003	beans	0.2 (0.5)	MRLs are for the sum of azocyclotin, cyhexatin and dicyclohexyltin oxide, expressed as cyhexatin. MRLs for azocyclotin and cyhexatin are not additive (see paras 4.3 and 4.15).
benomyl 1973, 1975, 1978	*0.02			ADIs have been estimated for benomyl and carbendazim. MRLs for carbendazim apply also to benomyl.
bitertanol	*0.005 (1987)			Residue data to be evaluated in 1984.
bromide ion 1968, 1969, 1971, 1979, 1981	1	*celery	300	
		*cucumbers	50	
		*tomatoes	75	
carbendazim 1973, 1976, 1977, 1978	*0.01	apricots	10 (5)	As an ADI has been estimated, previously recorded GLs are replaced by MRLs. Only new and amended limits are listed here. They are for carbendazim. Metabolites are not included.
		*asparagus 1/	0.1**	
		*blueberries	5	
		*chestnuts 1/	0.2	
		*eggplant	0.5	
		grapes	10 (5)	
		*hops, dry	50	
		melons	2 (0.5)	
		*oats	0.5	
		*onions	2	
		*peppers	5	1/ MRLs based on data from treatment with benomyl.
		*pineapples 1/	20	
		*rutabagas	0.1**	
		*soybeans	0.2	

Pesticide and year of previous evaluations	ADI (mg/kg body weight)	Commodity	Recommended MRL (mg/kg)	Remarks
carbendazim (contd)		*sweet potatoes 1/	1	
		*taro 1/	0.1**	
		bean fodder	50 (30 for bean vines)	
		peanut hay	5 (2)	
		peanut hulls 1/	1 (0.2)	
		rice straw	15 (2)	
		*soybean hay	0.1**	
		sugarbeet tops	10 (5)	
		wheat straw 1/	5 (2)	
carbophenothion 1972, 1976, 1977, 1979, 1980	0.0005			The definition of the residue in commodities of plant origin is changed, and is now identical to that in commodities of animal origin: sum of carbophenothion, its sulphoxide and its sulphone, expressed as carbophenothion.
chlorothalonil 1974, 1977, 1978, 1979, 1981	0.005 (1985)	lima beans (without pod)	0.5 (0.5 for lima beans)	Temporary ADI extended at same level. The definition of the residue is changed to chlorothalonil. Metabolites are not included. The temporary MRL for grapes (5 mg/kg) remains temporary, irrespective of the status of the ADI, pending further residues data (see para 4.11).
chlorpyrifos 1972, 1974, 1975, 1977, 1981, 1982	0.01	*dried vine fruit	2	MRLs are for chlorpyrifos. Metabolites are not included.
cyanofenphos 1975, 1978, 1980, 1982	No ADI			Temporary ADI withdrawn. Temporary MRLs replaced by GLs.
cyhexatin 1970, 1973, 1974, 1975, 1977, 1978, 1980, 1981, 1982	0.008	beans	0.2 (0.5)	MRLs are for the sum of cyhexatin and dicyclohexyltin oxide, expressed as cyhexatin. MRLs for cyhexatin and azocyclotin are not additive (see paras 4.3 and 4.15).
		*Kiwi fruit	5	
		peaches	5 (2)	

Pesticide and year of previous evaluations	ADI (mg/kg body weight)	Commodity	Recommended MRL (mg/kg)	Remarks
dichlofluanid 1969, 1974, 1977, 1979, 1981, 1982	0.3			Temporary ADI replaced by ADI at same level. Temporary MRLs replaced by MRLs.
diflubenzuron 1981	0.004 (1984)	mushrooms plums	0.1 (0.2) MRL (1) withdrawn	Temporary MRLs are for diflubenzuron. Metabolites are not included.
ethiofencarb 1977, 1978, 1981, 1982	0.1	beans (with pod)	2 (0.5)	MRLs are for the sum of ethiofencarb, its sulphoxide and its sulphone.
ethion 1968, 1969, 1970, 1972, 1975, 1982	0.001 (1985)			The definition of the residue is changed to ethion. Metabolites are not included.
fenchlorphos 1968, 1972	0.01			The definition of the residue in commodities of plant origin is changed to fenchlorphos and is now identical to that in commodities of animal origin.
fenitrothion 1969, 1974, 1976, 1977, 1979, 1982	0.001 (1984)	oranges *mandarins	2 (0.2) 2	The definition of the residue is changed to fenitrothion. Metabolites are not included. The MRLs for oranges and mandarins remain temporary, irrespective of the status of the ADI, pending the receipt of information on GAP.
mecarbam 1980	0.001 (1985)	*citrus fruit	2	Temporary ADI extended at same level. Temporary MRL replaces the temporary MRL for oranges at the same level. It is for mecarbam. Metabolites are not included.
methiocarb 1981	0.001	*Chinese radishes citrus fruit *sorghum strawberries sugarbeets	5 0.05**(0.02) 5 0.05** (0.02**) 0.05** (0.02)	Previously estimated ADI confirmed. MRLs are for the sum of methiocarb, its sulphoxide and its sulphone.
		apples blueberries cherries currants (red) grapes peaches	MRL withdrawn (10) " " (25) " " (10) " " (5) " " (5) " " (15)	MRLs withdrawn pending toxicological re-evaluation by JMPR (see report, para. 4.33).

Pesticide and year of previous evaluations	ADI (mg/kg body weight)	Commodity	Recommended MRL (mg/kg)	Remarks
methiocarb (contd)		*eggs *meat *milk *poultry	0.05** 1/ 0.05** 1/ 0.05** 1/ 0.05** 1/	1/ MRLs do not allow for possible feeding of straw and similar fodder from crops sprayed with methiocarb to repel birds.
oxamyl 1980	0.01 (1984,FAO) (1985,WHO)	bananas *coffee beans cucumbers *onions soybeans (dry) *sugarcane	0.2 (0.05**) 0.1 0.5 (2) 0.05** 0.1 (0.05**) 0.05**	Temporary ADI extended at same level. Previously recommended MRLs for maize and soybeans were temporary, irrespective of the status of the ADI, pending the receipt of data on residues and use patterns, which have now been supplied. Temporary MRLs are for the sum of oxamyl and its oxime, expressed as oxamyl.
permethrin 1979, 1980, 1981, 1982	0.05	beans (with pod) *legume oilseeds *sorghum fodder	1 (0.5 for beans, whole green) 0.1 20 (on a dry weight basis)	MRLs are for permethrin (sum of isomers). Metabolites are not included.
2-phenylphenol 1969, 1975	0.02 (1985)			ADI replaced by temporary ADI at lower level. MRLs replaced by temporary MRLs.
phorate 1977, 1982	0.0002 (1985)			Temporary ADI extended at same level.
phoxim 1982	0.0005 (1984)	*beans *cauliflower *cereal grains *cottonseed *lettuce *potatoes *tomatoes *carcase meat of cattle *carcase meat of sheep	0.05** 0.05** 0.05** 0.05** 0.1 0.05** 0.2 0.2 (in the carcase fat) 1 (in the carcase fat)	MRLs are for phoxim. Metabolites are not included.

Pesticide and year of previous evaluations	ADI (mg/kg body weight)	Commodity	Recommended MRL (mg/kg)		Remarks
pirimiphos-methyl 1974, 1976, 1977, 1979	0.01				The definition of the residue in commodities of plant origin is changed to pirimiphos-methyl and is now identical to that in commodities of animal origin.
prochloraz	*0.01	*barley	0.05**		MRLs are for prochloraz. Metabolites are not included. The MRL for citrus is temporary until information on GAP is provided.
		*citrus fruit	5		
		*mushrooms	2		
		*oats	0.05**		
		*rye	0.05**		
		*wheat	0.05**		
		*barley straw	0.2		
		*oat straw	0.2		
		*rye straw	0.2		
		*wheat straw	0.2		
		*avocados	5		MRLs are for the sum of prochloraz and its free and conjugated metabolites containing the 2,4,6-tri=chlorophenol moiety. They are all temporary until information on GAP and/or trials data in which residues are determined as prochloraz have been provided.
		*bananas	5		
		*mangos	2		
		*papayas	1		
		*rapeseed	0.5		
		*stone fruit	1		
propineb 1977, 1980	0.005 (1985)				Temporary ADI extended at same level.
propoxur 1973, 1977, 1981	0.02	cereal grains	0.1	(0.5)	The definition of the residue is changed to propoxur. Metabolites are not included. The amended MRLs have been lowered to accommodate the revised residue definition.
		fodders and straws (green)	1	(5)	
		legume animal feeds (green)	1	(5)	
tecnazene 1974, 1978, 1981	0.01				Temporary ADI replaced by ADI at same level. Temporary MRLs replaced by MRLs.
thiram 1965, 1967, 1970, 1974, 1977, 1980	0.005 (1985)				Temporary ADI extended at same level.

Pesticide and year of previous evaluations	ADI (mg/kg body weight)	Commodity	Recommended MRL (mg/kg)	Remarks
triadimefon 1979, 1981	0.01 (1985)	*pineapple	2	Temporary ADI extended at same level. MRLs are for the sum of triadimefon and "triadimenol".
		*pome fruit	0.5	
		wheat	0.2 (0.1**)	
		*fodder beet	0.1**	
		*fodder beet leaves	0.1**	
		*poultry meat	0.1**	
triazophos 1982	0.0002 (1986)	*bananas	2	Temporary MRLs are for triazophos. Metabolites are not included.
		*Brussels sprouts	0.1	
		*cabbage	0.1	
		*carrots	0.1	
		*cereal grains	0.5**	
		*citrus fruit	2	
		*coffee beans	0.05**	
		*cottonseed	0.1	
		*onions	0.1	
		*pome fruit	0.2	
		*peas	0.1	
		*potatoes	0.5**	
		*sugarbeet	0.05**	

PART 2. - GUIDELINE LEVELS (GLs) (NO ACCEPTABLE DAILY INTAKES)

Pesticide and year of previous evaluations	Commodity	GL (mg/kg)	Remarks
azinphos-ethyl 1973			The definition of the residue is changed to azinphos-ethyl. Metabolites are not included.
butocarboxim	*apples	2	GLs are for the sum of butocarboxim, its sulphoxide and its sulphone, expressed as butocarboxim.
	*beans (with pod)	5	
	*cabbage	1	
	*citrus fruit	2	
	*lettuce	2	The GL for lettuce is based exclusively on data from trials on outdoor crops.
	*radishes	2	
	*stone fruit	5	
	*sugarbeet	0.1	
	*tomatoes	1	
	*sugarbeet leaves	5	
nitrofen	*wheat	0.01**	GLs are for nitrofen. Metabolites are not included.
	*wheat straw.	0.05	

ANNEX IIA

NEW MAXIMUM RESIDUE LIMITS RECOMMENDED AT THE 1983 MEETING CLASSIFIED IN COMMODITY GROUPS

A01.0100 ROOT AND TUBER VEGETABLES

.0106	beets (fodder)	triadimefon (0.1**)
.0109	carrots	triazophos (0.1)
.0128	potatoes	phoxim (0.05**), triazophos (0.05**)
.0129	Chinese radishes	methiocarb (5)
.0117	taro	carbendazim (0.1**)
.0131	rutabagas	carbendazim (0.1**)
.0136	sugarbeets	triazophos (0.05**)
.0137	sweet potatoes	carbendazim (1)

A01.0200 BULB VEGETABLES

.0203	onions	carbendazim (2), oxamyl (0.05**), triazophos (0.1)

A01.0300 LEAFY VEGETABLES

.0327	lettuce	phoxim (0.1)
	fodder beet leaves	triadimefon (0.1**)

A01.0400 BRASSICA LEAFY VEGETABLES

.0403	Brussels sprouts	triazophos (0.1)
.0404	cabbage	triazophos (0.1)
.0407	cauliflower	phoxim (0.05**)

A01.0500 STEM VEGETABLES

.0502	asparagus	carbendazim (0.1**)
.0505	celery	bromide ion (300)

A01.0600 LEGUME VEGETABLES

.0601	beans	phoxim (0.05**)
.0614	soybeans	carbendazim (0.2)
.0620	peas	triazophos (0.1)

A01.0700 FRUITING VEGETABLES - EDIBLE PEEL

.0705	cucumbers	bromide ion (50)
.0706	eggplants	carbendazim (0.5)
.0710	peppers	carbendazim (5)
.0713	tomatoes	bromide ion (75), phoxim (0.2)

A02.0900 CITRUS FRUIT

mecarbam (2), prochloraz (5), triazophos (2)

.0909	mandarins	fenitrothion (2)

A02.1000 POME FRUIT

triadimefon (0.5), triazophos (0.2)

A02.1100 STONE FRUIT

prochloraz (1)

A02.1200 SMALL FRUITS AND BERRIES

.1203	blueberries	carbendazim (5)

A02.1400 ASSORTED FRUITS - INEDIBLE PEEL

.1402	avocados	prochloraz (5)
.1403	bananas	prochloraz (5), triazophos (2)
.1419	Kiwi fruit	cyhexatin (5)
.1423	mangos	prochloraz (2)
.1428	papayas	prochloraz (1)
.1432	pineapples	carbendazim (20), triadimefon (2)

A03.1500 CEREAL GRAINS

phoxim (0.05**), triazophos (0.05**)

.1501	barley	prochloraz (0.05**)
.1510	oats	carbendazim (0.5), prochloraz (0.05**)
.1516	rye	prochloraz (0.05**)
.1517	sorghum	methiocarb (5)
.1521	wheat	prochloraz (0.05**)

A03.1600 FODDERS AND STRAWS

.1601	barley straw	prochloraz (0.2)
.1607	oat straw	prochloraz (0.2)
.1609	rye straw	prochloraz (0.2)
.1610	sorghum fodder	permethrin (20)
.1612	wheat straw	prochloraz (0.2)

A04.1700 LEGUME OILSEED

permethrin (0.1)

A04.1800 LEGUME ANIMAL FEEDS

.1811	soybean hay	carbendazim (0.1**)

A05.1900 NUTS AND SEEDS

.1906	chestnuts	carbendazim (0.2)

A05.2000 OIL SEED

.2004	cottonseed	phoxim (0.05**), triazophos (0.1)
.2011	rapeseed	prochloraz (0.5)

A05.2100 TROPICAL SEED

.2102	coffee beans	oxamyl (0.1), triazophos (0.05**)

MISCELLANEOUS CROPS

mushrooms	prochloraz (2)
hops (dried)	carbendazim (50)
sugarcane	oxamyl (0.05**)

ANIMAL PRODUCTS

B07.2500 CARCASE MEAT

methiocarb (0.05**)

.2503	carcase meat of cattle	phoxim (0.2 in the fat)
.2509	carcase meat of sheep	phoxim (1 in the fat)

B07.2800 MILK

methiocarb (0.05**)

B08.3000 POULTRY MEATS

methiocarb (0.05**), triadimefon (0.1**)

B08.3300 EGGS

methiocarb (0.05**)

PROCESSED FOOD AND FEEDS

dried vine fruit	chlorpyrifos (2)

ANNEX IIB

CHANGES IN MAXIMUM RESIDUES LIMITS RECOMMENDED AT THE 1983 MEETING

		from	to
amitraz	carcase meat of sheep	0.2	0.1
azocyclotin	beans	0.5	0.2
benomyl and carbendazim	apricots	5	10
	grapes	5	10
	melons	0.5	2
	bean fodder	30	50
	peanut hay	2	5
	peanut hulls	0.2	1
	rice straw	2	15
	sugarbeet tops	5	10
	wheat straw	2	5
chlorthalonil	lima beans (without pod)	0.5 (lima beans)	0.5
cyhexatin	beans	0.5	0.2
	peaches	2	5
diflubenzuron	mushrooms	0.2	0.1
	plums	1	withdrawn
ethiofencarb	beans (with pod)	0.5	2
fenitrothion	oranges	0.2	2
methiocarb	citrus fruit	0.02	0.05**
	strawberries	0.02**	0.05**
	sugarbeet	0.02	0.05**
	apples	10	withdrawn
	blueberries	25	"
	cherries	10	"
	currant (red)	5	"
	grapes	5	"
	peaches	15	"
oxamyl	bananas	0.05**	0.2
	cucumbers	2	0.5
	soybeans (dry)	(0.05**)	0.1
permethrin	beans (with pod)	0.5 (beans whole green)	1
propoxur	cereal grains	0.5	0.1
	fodders and straws (green)	5	1
	legume animal feeds (green)	5	1
triadimefon	wheat	(0.1**)	0.2

ANNEX III

FAO/WHO JOINT MEETING ON PESTICIDE RESIDUES
INDEX OF REPORTS AND EVALUATIONS

ACEPHATE	1976, 1979, 1981, 1982
ALDICARB	1979, 1982
ALDRIN	1965, 1966, 1967, 1974, 1975, 1977
ALLETHRIN	1965
AMINOCARB	1978, 1979
AMITRAZ	1980, 1983
AMITROLE	1974, 1977
AZINPHOS-ETHYL	1973, 1983
AZINPHOS-METHYL	1965, 1968, 1972, 1973, 1974
AZOCYCLOTIN	1979, 1981, 1982, 1983
BENDIOCARB	1982
BENOMYL	1973, 1975, 1983
BHC (technical) (see also lindane)	1965, 1968, 1973
BINAPACRYL	1969, 1974, 1982
BIORESMETHRIN	1975, 1976
BITERTANOL	1983
BROMIDE ION	1968, 1969, 1971, 1979, 1981, 1983
BROMOMETHANE	1966, 1967, 1968, 1971, 1979
BROMOPHOS	1972, 1975, 1977, 1982
BROMOPHOS-ETHYL	1972, 1975, 1977
BROMOPROPYLATE	1973
BUTOCARBOXIM	1983
sec-BUTYLAMINE	1975, 1977, 1978, 1979, 1980, 1981
CAMPHECHLOR	1968, 1973
CAPTAFOL	1969, 1973, 1974, 1976, 1977, 1982
CAPTAN	1965, 1969, 1973, 1974, 1977, 1978, 1980, 1982
CARBARYL	1965, 1966, 1967 1968, 1969, 1970, 1973, 1975, 1976, 1977, 1979
CARBENDAZIM	1973, 1976, 1977, 1978, 1983
CARBOFURAN	1976, 1979, 1980, 1982
CARBON DISULPHIDE	1965, 1967, 1968, 1971
CARBON TETRACHLORIDE	1965, 1967, 1968, 1971, 1979
CARBOPHENOTHION	1972, 1976, 1977, 1979, 1980, 1983
CARTAP	1976, 1978
CHINOMETHIONAT	1968, 1974, 1977, 1981, 1983
CHLORBENSIDE	1965
CHLORDANE	1965, 1967, 1969, 1970, 1972, 1974, 1977, 1982
CHLORDIMEFORM	1971, 1975, 1977, 1978, 1979, 1980
CHLORFENSON	1965
CHLORFENVINPHOS	1971
CHLORMEQUAT	1970, 1972, 1976
CHLOROBENZILATE	1965, 1968, 1972, 1975, 1977, 1980
CHLOROPROPYLATE	1968, 1972

CHLOROTHALONIL	1974, 1977, 1978, 1979, 1981, 1983
CHLORPROPHAM	1965
CHLORPYRIFOS	1972, 1974, 1975, 1977, 1981, 1982, 1983
CHLORPYRIFOS-METHYL	1975, 1979
CHLORTHION	1965
COUMAPHOS	1968, 1972, 1975, 1978, 1983
CRUFOMATE	1968, 1972
CYANOFENPHOS	1975, 1978, 1980, 1982, 1983
CYHEXATIN (TRICYCLOHEXYL=TIN HYDROXIDE '70 & '73)	1970, 1973, 1974, 1975, 1977, 1978, 1980, 1981, 1982, 1983
CYPERMETHRIN	1979, 1981, 1982, 1983
2,4-D	1970, 1971, 1974, 1975, 1980
DAMINOZIDE	1977, 1983
DDT	1965, 1966, 1967, 1968, 1969, 1979, 1980, 1983
DELTAMETHRIN	1980, 1981, 1982
DEMETON	1965, 1967, 1975, 1982, 1983
DEMETON-S-METHYL and related compounds	1965, 1967, 1968, 1973, 1979, 1982, 1983
DIALIFOS	1976, 1982
DIAZINON	1965, 1966, 1967, 1968, 1970, 1975, 1979
1,2 DIBROMO-ETHANE	1966, 1967, 1968, 1971, 1979
DICHLOFLUANID	1969, 1974, 1977, 1979, 1981, 1982, 1983
1,2-DICHLORO=ETHANE	1967, 1971, 1979
DICHLORVOS	1965, 1966, 1967, 1969, 1970, 1974, 1977
DICLORAN	1974, 1977
DICOFOL	1968, 1970, 1974
DIELDRIN	1965, 1966, 1967, 1968, 1969, 1970, 1974, 1977
DIFLUBENZURON	1981, 1983
DIMETHOATE	1965, 1966, 1967, 1970, 1977, 1983
DIMETHRIN	1965
DINOCAP	1969, 1974
DIOXATHION	1968, 1972
DIPHENYL	1966, 1967
DIPHENYLAMINE	1969, 1976, 1979, 1982
DIQUAT	1970, 1972, 1976, 1977, 1978
DISULFOTON	1973, 1975, 1979, 1981
DITHIOCARBAMATE FUNGICIDES	1965, 1967, 1970, 1974, 1977, 1980, 1983
DNOC	1965
DODINE	1974, 1976, 1977
EDIFENPHOS	1976, 1979, 1981
ENDOSULFAN	1965, 1967, 1968, 1971, 1974, 1975, 1982
ENDRIN	1965, 1970, 1974, 1975
ETHEPHON	1977, 1978, 1983
ETHIOFENCARB	1977, 1978, 1981, 1982, 1983
ETHION	1968, 1969, 1970, 1972, 1975, 1982, 1983

ETHOPROPHOS	1983
ETHOXYQUIN	1969
ETHYLENE OXIDE	1965, 1968, 1971
ETRIMFOS	1980, 1982
FENAMIPHOS	1974, 1977, 1978, 1980
FENCHLORPHOS	1968, 1972, 1980, 1983
FENITROTHION	1969, 1974, 1976, 1977, 1979, 1982, 1983
FENSULFOTHION	1972, 1982, 1983
FENTHION	1971, 1975, 1977, 1978, 1979, 1980, 1983
FENTIN compounds	1965, 1970, 1972
FENVALERATE	1979, 1981, 1982
FERBAM	see dithiocarbamate fungicides
FOLPET	1969, 1973, 1974, 1982
FORMOTHION	1969, 1972, 1973
FENBUTATIN OXIDE	1977, 1979
GUAZATINE	1978, 1980
HEPTACHLOR	1965, 1966, 1967, 1968, 1969, 1970, 1974, 1975, 1977
HEXACHLORO=BENZENE	1969, 1973, 1974, 1978
HYDROGEN CYANIDE	1965
HYDROGEN PHOSPHIDE	1965, 1966, 1967, 1969, 1971
IMAZALIL	1977, 1980
IPRODIONE	1977, 1980
ISOFENPHOS	1981, 1982
LEAD ARSENATE	1965, 1968
LEPTOPHOS	1974, 1975, 1978
LINDANE	1965, 1966, 1967, 1968, 1969, 1973, 1974, 1975, 1977, 1978, 1979
MALATHION	1965, 1966, 1967, 1968, 1969, 1970, 1973, 1975, 1977
MALEIC HYDRAZIDE	1976, 1977, 1980
MANEB	see dithiocarbamate fungicides
MANCOZEB	1967, 1970, 1974, 1977, 1980
MECARBAM	1980, 1983
METALAXYL	1982
METHACRIFOS	1980, 1982
METHAMIDOPHOS	1976, 1979, 1981, 1982
METHIDATHION	1972, 1975, 1979
METHIOCARB	1981, 1983
METHOMYL	1975, 1976, 1977, 1978
METHOXYCHLOR	1965, 1977
METHYL BROMIDE	see bromomethane
MEVINPHOS	1965, 1972
MGK 264	1967
MONOCROTOPHOS	1972, 1975
NABAM	see dithiocarbamate fungicides
NITROFEN	1983
OMETHOATE	1971, 1975, 1978, 1979, 1981
ORGANOMERCURY	1965, 1966, 1967
OXAMYL	1980, 1983
OXYDEMETON METHYL	1965, 1967, 1968, 1973